# Building Constructions

# Building Constructions

Ramesh Shah

RANDOM

RANDOM PUBLICATIONS

NEW DELHI (INDIA)

**Building Constructions**

---

ISBN 978-93-5111-435-2



Published in 2014 in India by

**RANDOM PUBLICATIONS**

4376-A/4B, Gali Murari Lal, Ansari Road
New Delhi-110 002
Phone : +91-11-43580356, +91-11-23289044
e-mail: randomexports@gmail.com, sales@randompublications.com,
info@randompublications.com

Reprinted 2025

*Type Setting by:* Friends Media, Delhi-110089
*Digitally Printed at:* Replika Press Pvt. Ltd.

# Preface

Building construction, the techniques and industry involved in the assembly and erection of structures, primarily those used to provide shelter. Building construction is an ancient human activity. It began with the purely functional need for a controlled environment to moderate the effects of climate. Constructed shelters were one means by which human beings were able to adapt themselves to a wide variety of climates and become a global species. Human shelters were at first very simple and perhaps lasted only a few days or months. Over time, however, even temporary structures evolved into such highly refined forms as the igloo. Gradually more durable structures began to appear, particularly after the advent of agriculture, when people began to stay in one place for long periods. The first shelters were dwellings, but later other functions, such as food storage and ceremony, were housed in separate buildings. Some structures began to have symbolic as well as functional value, marking the beginning of the distinction between architecture and building. The history of building is marked by a number of trends. One is the increasing durability of the materials used. Early building materials were perishable, such as leaves, branches, and animal hides. Later, more durable natural materials—such as clay, stone, and timber—and, finally, synthetic materials—such as brick, concrete, metals, and plastics—were used. Another is a quest for buildings of ever greater height and span; this was made possible by the development of stronger materials and by knowledge of how materials behave and how to exploit them to greater advantage. A third major trend involves the degree of control exercised over the interior environment of buildings: increasingly precise regulation of air temperature, light and sound levels, humidity, odours, air speed, and other factors that affect human comfort has been possible. Yet another trend is the change in energy available to the construction process, starting with human muscle power and developing toward the powerful machinery used today.

The present state of building construction is complex. There is a wide range of building products and systems which are aimed primarily at groups of building types or markets. The design process for buildings is highly organized and draws upon research establishments that study material properties and performance, code officials who adopt and enforce safety standards, and design professionals who determine user needs and design a building to meet those needs. The construction process is also highly organized; it includes the manufacturers of building products and systems, the craftsmen who assemble them on the building site, the contractors who employ and coordinate the work of the craftsmen, and consultants who specialize in such aspects as construction management, quality control, and insurance.

This book offers a comprehensive description of the applications of various fields in this subject. The book will be appropriate as a guide for students.

I would like to thank my team for standing beside me throughout my career and writing this book. My special thanks go to "Random Publications" who have published the book.

*—Ramesh Shah*

# Contents

# 1

# Construction

In the fields of architecture and civil engineering, construction is a process that consists of the building or assembling of infrastructure. Far from being a single activity, large scale construction is a feat of human multitasking. Normally, the job is managed by a project manager, and supervised by a construction manager, design engineer, construction engineer or project architect.

For the successful execution of a project, effective planning is essential. Involved with the design and execution of the infrastructure in question must consider the environmental impact of the job, the successful scheduling, budgeting, construction site safety, availability of building materials, logistics, inconvenience to the public caused by construction delays and bidding, etc.

## Types of Construction Projects

In general, there are four types of construction:

1. Residential Building construction
2. Industrial construction
3. Commercial Building construction
4. Heavy Civil construction

Each type of construction project requires a unique team to plan, design, construct and maintain the project.

## Building Construction

Building construction is the process of adding structure to real property. The vast majority of building construction jobs are small renovations, such as addition of a room, or renovation of a bathroom.

Often, the owner of the property acts as labourer, paymaster, and design team for the entire project. However, all building construction projects include some elements in common – design, financial, estimating and legal considerations. Many projects of varying sizes reach undesirable end results, such as structural collapse, cost overruns, and/or litigation. For this reason, those with experience in the field make detailed plans and maintain careful oversight during the project to ensure a positive outcome.

Commercial building construction is procured privately or publicly utilising various delivery methodologies, including cost estimating, hard bid, negotiated price, traditional, management contracting, construction management-at-risk, design & build and design-build bridging.

Residential construction practices, technologies, and resources must conform to local building authority regulations and codes of practice. Materials readily available in the area generally dictate the construction materials used (e.g. brick versus stone, versus timber). Cost of construction on a per square metre (or per square foot) basis for houses can vary dramatically based on site conditions, local regulations, economies of scale (custom designed homes are always more expensive to build) and the availability of skilled tradespeople. As residential construction (as well as all other types of construction) can generate a lot of waste, careful planning again is needed here.

The most popular method of residential construction in the United States is wood framed construction. As efficiency codes have come into effect in recent years, new construction technologies and methods have emerged. University Construction Management departments are on the cutting edge of the newest methods of construction intended to improve efficiency, performance and reduce construction waste.

New techniques of building construction are being researched, made possible by advances in 3D printing technology. In a form of *additive building construction,* similar to the additive manufacturing techniques for manufactured parts, building printing is making it possible to flexibly construct small commercial buildings and private habitations in around 20 hours, with built-in plumbing and electrical facilities, in one continuous build, using large 3D printers. Working versions of 3D-printing building technology are already printing 2 metres (6 ft 7 in) of building material per hour as of January 2013, with the next-generation printers capable of 3.5 metres (11 ft) per hour, sufficient to complete a building in a week. Dutch architect

Janjaap Ruijssenaars's performative architecture 3D-printed building is scheduled to be built in 2014.

## Construction Processes

### *Design Team*

In the modern industrialized world, construction usually involves the translation of designs into reality. A formal design team may be assembled to plan the physical proceedings, and to integrate those proceedings with the other parts.

***Figure:*** *Shasta Dam under construction in June **1942***

The design usually consists of drawings and specifications, usually prepared by a design team including surveyors, civil engineers, cost engineers (or quantity surveyors), mechanical engineers, electrical engineers, structural engineers, fire protection engineers, planning consultants, architectural consultants, and archaeological consultants. The design team is most commonly employed by (i.e. in contract with) the property owner. Under this system, once the design is completed by the design team, a number of construction companies or construction management companies may then be asked to make a bid for the work, either based directly on the design, or on the basis of drawings and a bill of quantities provided by a quantity surveyor. Following evaluation of bids, the owner will typically award a contract to the most cost efficient bidder.

The modern trend in design is toward integration of previously separated specialities, especially among large firms. In the past, architects, interior designers, engineers, developers, construction managers, and general contractors were more likely to be entirely separate companies, even in the larger firms. Presently, a firm that is nominally an "architecture" or "construction management" firm may have experts from all related fields as employees, or to have an associated company that provides each necessary skill. Thus, each such firm may offer itself as "one-stop shopping" for a construction project, from beginning to end. This is designated as a "design Build" contract where the contractor is given a performance specification and must undertake the project from design to construction, while adhering to the performance specifications.

Several project structures can assist the owner in this integration, including design-build, partnering and construction management. In general, each of these project structures allows the owner to integrate the services of architects, interior designers, engineers and constructors throughout design and construction. In response, many companies are growing beyond traditional offerings of design or construction services alone and are placing more emphasis on establishing relationships with other necessary participants through the design-build process.

The increasing complexity of construction projects creates the need for design professionals trained in all phases of the project's life-cycle and develop an appreciation of the building as an advanced technological system requiring close integration of many sub-systems and their individual components, including sustainability. Building engineering is an emerging discipline that attempts to meet this new challenge.

### Financial Advisors

Construction projects can suffer from preventable financial problems. Underbids ask for too little money to complete the project. Cash flow problems exist when the present amount of funding cannot cover the current costs for labour and materials, and because they are a matter of having sufficient funds at a specific time, can arise even when the overall total is enough. Fraud is a problem in many fields, but is notoriously prevalent in the construction field. Financial planning for the project is intended to ensure that a solid plan with adequate safeguards and contingency plans are in place before the project is started and is required to ensure that the plan is properly executed over the life of the project.

Mortgage bankers, accountants, and cost engineers are likely participants in creating an overall plan for the financial management of the building construction project. The presence of the mortgage banker is highly likely, even in relatively small projects since the owner's equity in the property is the most obvious source of funding for a building project. Accountants act to study the expected monetary flow over the life of the project and to monitor the payouts throughout the process. Cost engineers and estimators apply expertise to relate the work and materials involved to a proper valuation. Cost overruns with government projects have occurred when the contractor was able to identify change orders or changes in the project resulting in large increases in cost, which are not subject to competition by other firm as they have already been eliminated from consideration after the initial bid.

Large projects can involve highly complex financial plans and often start with a conceptual estimate performed by a building estimator. As portions of a project are completed, they may be sold, supplanting one lender or owner for another, while the logistical requirements of having the right trades and materials available for each stage of the building construction project carries forward. In many English-speaking countries, but not the United States, projects typically use quantity surveyors.

### Legal Aspects

A construction project must fit into the legal framework governing the property. These include governmental regulations on the use of property, and obligations that are created in the process of construction.

The project must adhere to zoning and building code requirements. Constructing a project that fails to adhere to codes will not benefit the owner. Some legal requirements come from malum in se considerations, or the desire to prevent things that are indisputably bad – bridge collapses or explosions. Other legal requirements come from malumprohibitum considerations, or things that are a matter of custom or expectation, such as isolating businesses to a business district and residences to a residential district. An attorney may seek changes or exemptions in the law governing the land where the building will be built, either by arguing that a rule is inapplicable (the bridge design will not collapse), or that the custom is no longer needed (acceptance of live-work spaces has grown in the community).

A construction project is a complex net of contracts and other legal obligations, each of which must be carefully considered. A contract is the exchange of a set of obligations between two or more parties, but it is not so simple a matter as trying to get the other side to agree to as much as possible in exchange for as little as possible. The time element in construction means that a delay costs money, and in cases of bottlenecks, the delay can be extremely expensive. Thus, the contracts must be designed to ensure that each side is capable of performing the obligations set out. Contracts that set out clear expectations and clear paths to accomplishing those expectations are far more likely to result in the project flowing smoothly, whereas poorly drafted contracts lead to confusion and collapse.

Legal advisors in the beginning of a construction project seek to identify ambiguities and other potential sources of trouble in the contract structure, and to present options for preventing problems. Throughout the process of the project, they work to avoid and resolve conflicts that arise. In each case, the lawyer facilitates an exchange of obligations that matches the reality of the project.

### Interaction of Expertise

Design, finance, and legal aspects overlap and interrelate. The design must be not only structurally sound and appropriate for the use and location, but must also be financially possible to build, and legal to use. The financial structure must accommodate the need for building the design provided, and must pay amounts that are legally owed. The legal structure must integrate the design into the surrounding legal framework, and enforce the financial consequences of the construction process.

### Procurement

Procurement describes the merging of activities undertaken by the client to obtain a building. There are many different methods of construction procurement; however the three most common types of procurement are:

1. Traditional (Design-bid-build)
2. Design and build
3. Management contracting

There is also a growing number of new forms of procurement that involve relationship contracting where the emphasis is on a co-operative relationship between the principal and contractor and other stakeholders within a construction project. New forms include partnering such as Public-Private Partnering (PPPs) aka private finance initiatives (PFIs) and alliances such as "pure" or "project" alliances and "impure" or "strategic" alliances. The focus on co-operation is to ameliorate the many problems that arise from the often highly competitive and adversarial practices within the construction industry.

### Traditional

This is the most common method of construction procurement and is well established and recognised. In this arrangement, the architect or engineer acts as the project coordinator. His or her role is to design the works, prepare the specifications and produce construction drawings, administer the contract, tender the works, and manage the works from inception to completion. There are direct contractual links between the architect's client and the main contractor. Any subcontractor will have a direct contractual relationship with the main contractor.

### Design and Build

This approach has become more common in recent years and involves the client contracting a single entity to both provide a design and to build that design. In some cases, the Design and Build (D & B) package can also include finding the site, arranging funding and applying for all necessary statutory consents.

The owner produces a list of requirements for a project, giving an overall view of the project's goals. Several D&B contractors present different ideas about how to accomplish these goals. The owner selects the ideas he or she likes best and hires the appropriate contractor. Often, it is not just one contractor, but a consortium of several contractors working together. Once a contractor (or a consortium/ consortia) has been hired, they begin building the first phase of the project. As they build phase 1, they design phase 2. This is in contrast to a design-bid-build contract, where the project is completely designed by the owner, then bid on, then completed.

Kent Hansen pointed out that state departments of transportation (DOTs) usually use design build contracts as a way of getting projects done when states don't have the resources. In DOTs, design build contracts are usually used for very large projects.

## Management Procurement Systems

In this arrangement the client plays an active role in the procurement system by entering into separate contracts with the designer (architect or engineer), the construction manager, and individual trade contractors. The client takes on the contractual role, while the construction or project manager provides the active role of managing the separate trade contracts, and ensuring that they all work smoothly and effectively together.

Management procurement systems are often used to speed up the procurement processes, allow the client greater flexibility in design variation throughout the contract, the ability to appoint individual work contractors, separate contractual responsibility on each individual throughout the contract, and to provide greater client control.

### Authority Having Jurisdiction

In construction, the authority having jurisdiction (AHJ) is the governmental agency or sub-agency which regulates the construction process. In most cases, this is the municipality in which the building is located. However, construction performed for supra-municipal authorities are usually regulated directly by the owning authority, which becomes the AHJ.

Before the foundation can be dug, contractors are typically required to verify and have existing utility lines marked, either by the utilities themselves or through a company specialising in such services. This lessens the likelihood of damage to the existing electrical, water, sewage, phone, and cable facilities, which could cause outages and potentially hazardous situations. During the construction of a building, the municipal building inspector inspects the building periodically to ensure that the construction adheres to the approved plans and the local building code. Once construction is complete and a final inspection has been passed, an occupancy permit may be issued.

An operating building must remain in compliance with the fire code. The fire code is enforced by the local fire department.

Changes made to a building that affect safety, including its use, expansion, structural integrity, and fire protection items, usually require approval of the AHJ for review concerning the building code.

### Industry Characteristics

In the United States, the industry has around $850 billion in annual revenue according to statistics tracked by the Census Bureau, with an $857 billion annual rate in March 2013, of which $600 billion is private (split evenly between residential and nonresidential) and the remainder is government. As of 2005, there were about 667,000 firms employing 1 million contractors (200,000 general contractors, 38,000 heavy, and 432,000 speciality); the average contractor employed fewer than 10 employees. As a whole, the industry employed an estimated 5.8 million as of April 2013, with a 13.2% unemployment rate.

### *Safety*

Construction is one of the most dangerous occupations in the world, incurring more occupational fatalities than any other sector in both the United States and in the European Union. In 2009, the fatal occupational injury rate among construction workers in the United States was nearly three times that for all workers. Falls are one of the most common causes of fatal and non-fatal injuries among construction workers.

Proper safety equipment such as harnesses and guardrails and procedures such as securing ladders and inspecting scaffolding can curtail the risk of occupational injuries in the construction industry. Other major causes of fatalities in the construction industry include electrocution, transportation accidents, and trench cave-ins.

### *History*

The first huts and shelters were constructed by hand or with simple tools. As cities grew during the Bronze Age, a class of professional craftsmen, like bricklayers and carpenters, appeared. Occasionally, slaves were used for construction work. In the Middle Ages, these were organised into guilds.

In the 19th century, steam-powered machinery appeared, and later diesel- and electric powered vehicles such as cranes, excavators and bulldozers. Architecture and construction involves creating awesome structures that can show the beauty and creativity of the human intellect.

Fast-track construction has become more popular in the 21st century, with some estimates suggesting that 40% of projects.

## Building

In architecture, construction, engineering, real estate development and technology the term building refers to one of the following:

1. Any human-made structure used or intended for supporting or sheltering any use or continuous occupancy, or
2. An act of construction (i. e. the activity of building)

In this article, the first usage is generally intended unless otherwise specified. Buildings come in a wide amount of shapes and functions, and have been adapted throughout history for a wide number of factors, from building materials available, to weather conditions, to land prices, ground conditions, specific uses and aesthetic reasons.

Buildings serve several needs of society – primarily as shelter from weather and as general living space, to provide privacy, to store belongings and to comfortably live and work. A building as a shelter represents a physical division of the human habitat (a place of comfort and safety) and the *outside* (a place that at times may be harsh and harmful).

Ever since the first cave paintings, buildings have also become objects or canvasess of artistic expression. In recent years, interest in sustainable planning and building practices has also become part of the design process of many new buildings.

### Definitions

Building is defined in many aspects as:

- The act of constructing, erecting, or establishing.
- The art of constructing edifices, or the practice of civil architecture.
- That which is built; a fabric or edifice constructed, as a house, a church, castle, arena/ stadium, etc.
- The act of constructing or building something; "during the construction we had to take a detour"; "his hobby was the building of boats"
- The commercial activity involved in constructing buildings; "their main business is home construction"; "workers in the building trades"
- A structure that has a roof and walls and stands more or less permanently in one place; "there was a three-storey building on the corner"; "it was an imposing edifice"
- The occupants of a building; "the entire building complained about the noise"

To differentiate buildings in the usage of this article from other buildings and other structures that are *not* intended for continuous human occupancy, the latter are called non-building structures or simply structures.

*Structural height* in technical usage is the height to the highest architectural detail on building from street-level. Depending on how they are classified, spires and masts may or may not be included in this height. Spires and masts used as antennas are not generally included.

The definition of a *low-rise vs. a high-rise* building is a matter of debate, but generally three storeys or less is considered low-rise.

### History

The first shelter on Earth constructed by a relatively close ancestor to humans is believed to be built 500,000 years ago by an early ancestor of humans, Homo erectus.

### Types

***Residential:*** Residential buildings are called houses or homes, though buildings containing large numbers of separate dwelling units are often called apartment buildings or apartment blocks to differentiate them from 'individual' houses. Houses may also be built in pairs (semi-detached), in terraces where all but two of the houses have others either side; apartments may be built round courtyards or as rectangular blocks surrounded by a piece of ground of varying sizes. Houses which were built as a single dwelling may later be divided into apartments or bedsitters; they may also be converted to another use e.g. an office or a shop.

Building types may range from one-room wood-framed, masonry, or adobe dwellings to multi-million dollar high-rise buildings able to house thousands of people. Increasing settlement density in buildings (and smaller distances between buildings) is usually a response to high ground prices resulting from many people wanting to live close to work or similar attractors. Other common building materials are brick, concrete or combinations of either of these with stone.

### Multi-Storey

A Multi-Stored is a building that has multiple floors above ground in the building.

Multi-storey buildings aim to increase the floor area of the building without increasing the area of the land the building is built on, hence saving land and, in most cases, money (depending on material used and land prices in the area). The building with the most storeys is the BurjKhalifa, with 162.

### Creation

The practice of designing, constructing, and operating buildings is most usually a collective effort of different groups of professionals and trades. Depending on the size, complexity, and purpose of a particular building project, the project team may include:

- A real estate developer who secures funding for the project;
- One or more financial institutions or other investors that provide the funding

- Local planning and code authorities
- A Surveyor who performs an ALTA/ACSM and construction surveys throughout the project;
- Construction managers who coordinate the effort of different groups of project participants;
- Licensed architects and engineers who provide building design and prepare construction documents;
- Landscape architects;
- Interior designers;
- Other consultants;
- Contractors who provide construction services and install building systems such as climate control, electrical, plumbing, Decoration, fire protection, security and telecommunications;
- Marketing or leasing agents;
- Facility managers who are responsible for operating the building.

Regardless of their size or intended use, all buildings in the US must comply with zoning ordinances, building codes and other regulations such as fire codes, life safety codes and related standards. Vehicles—such as trailers, caravans, ships and passenger aircraft—are treated as "buildings" for life safety purposes.

### Ownership and Funding

- Mortgage loan
- Real estate developer

### Building Services

***Physical plant:*** Any building requires a certain amount of internal infrastructure to function, which includes such elements like heating / cooling, power and telecommunications, water and wastewater etc. Especially in commercial buildings (such as offices or factories), these can be extremely intricate systems taking up large amounts of space (sometimes located in separate areas or double floors / false ceilings) and constitute a big part of the regular maintenance required.

### Conveying Systems

Systems for transport of people within buildings:

- Elevator
- Escalator
- Moving sidewalk (horizontal and inclined)

Systems for transport of people between interconnected buildings:

- Skyway
- Underground city

### *Building Damage*

Buildings may be damaged during the construction of the building or during maintenance. There are several other reasons behind building damage like accidents such as storms, explosions and subsidence caused by mining or poor foundations. Buildings also may suffer from fire damage and flooding in special circumstances. They may also become dilapidated through lack of proper maintenance or alteration work improperly carried out.

## Home

A home is a place of residence and refuge. When it refers to a building, it is usually a place in which an individual or a family can live and store personal property, such as a single-family detached home or an apartment. It is generally a place to provide safety and is used as a centre from which people or animals base their daily activities. Most modern-day households contain sanitary facilities and a means of preparing food. Animals have their own homes as well, either living in the wild or shared with humans in a domesticated environment. Home might have a Sanskritaham (self) or akam (self) in Tamil as its etymology.

Home is also used to refer to the geographical area (be a suburb, town, city or country) in which a person grew up or feels they belong, or it can refer to the native habitat of a wild animal. There are cultures in which homes are mobile such as nomadic peoples. Sometimes, as an alternative to the definition of home as a physical locale ("Home is where you hang your hat"), home may be perceived to have no physical location. Instead, home may relate to a mental or emotional state of refuge or comfort. Popular sayings along these lines are "Home is where the heart is" or "You can never go home again".

### *Terminology*

The word "home" can be used for various types of residential community institutions in which people can live, such as nursing, retirement homes for seniors, prisons for criminals, treatment facilities, foster homes, etc. A home is generally a place that is close to the heart of the owner, and can become a prized possession.

In computer terminology, a 'home' may refer to a starting view that branches off into other tasks, e.g. a homepage or a desktop. In a full screen editor, home is often used to mean the top-leftmost character cell, or the leftmost cell on a line in a line editor. These are the initial ones used by left-to-right languages. A standard 101-key PC keyboard contains a Home key. Many home pages on the with introductory information, recent news or events, and links to subpages. "Home " may also refer to a home directory which contains the personal files of a given user of the computer system.

### *History*

Buildings such as huts and longhouses håve been used for living since the late Neolithic. Homes may be lost in many ways, such as Natural Disasters. Other people have lost their homes through events by humans such as Sherman's March to the Sea.

### *Psychological Impact*

Since it can be said that humans are generally creatures of habit, the state of a person's home has been known to physiologically influence their behavior, emotions, and overall mental health. The loss of a home (due to whatever reason, be it through accident or natural disaster, repossession, or in the case of children simply the decision to move on the part of the parents) can be a valid cause of relocation.

Some people may become homesick when they leave their home over an extended period of time. Sometimes homesickness can cause a person to feel actual symptoms of illness. It has been argued that psychologically "The strongest sense of home commonly coincides geographically with a dwelling. Usually the sense of home attenuates as one moves away from that point, but it does not do so in a fixed or regular way." Furthermore, places like homes can trigger self-reflection, thoughts about who someone is or used to be or who they might become. These types of reflections also occur in places where there is a collective historical identity, such as Gettysburg or Ground Zero.

## Tower Block

A tower block, high-rise, residential towers, apartment tower, office tower, apartment block, or block of flats, is a tall building or structure used as a residential and/or office building. In some areas they may be referred to as "MDU" standing for "Multi Dwelling Unit".

High-rise buildings became possible with the invention of the elevator (lift) and cheaper, more abundant building materials. The materials used for the structural system of high-rise buildings are reinforced concrete and steel. Most North American style skyscrapers have a steel frame, while residential blocks are usually constructed of concrete. There is no clear definition of any difference between a tower block and a skyscraper, although a building with fifty or more storeys is generally considered a skyscraper.

***Figure:*** *A high-rise residential apartment building in Hong Kong*

High-rise structures pose particular design challenges for structural and geotechnical engineers, particularly if situated in a seismically active region or if the underlying soils have geotechnical risk factors such as high compressibility or bay mud. They also pose serious challenges to firefighters during emergencies in high-rise structures. New and old building design, building systems like the building standpipe system, HVAC systems (heating, ventilation and air conditioning), fire sprinkler system and other things like stairwell and elevator evacuations pose significant problems. Studies are often required to ensure that pedestrian wind comfort and wind danger concerns are addressed.

Apartment blocks have technical and economic advantages in areas of high population density, and have become a distinctive feature of housing accommodation in virtually all densely populated urban areas around the world. In contrast with low-rise and single-family houses, apartment blocks accommodate more inhabitants per unit of area of land and decrease the cost of municipal infrastructure.

### Definition

Various bodies have defined "high-rise":

- Emporis Standards defines a high-rise as "A multi-story structure between 35–100 metres tall, or a building of unknown height from 12–39 floors."
- According to the building code of Hyderabad, India, a high-rise building is one with four floors or more, or one 15 metres or more in height.
- The *New Shorter Oxford English Dictionary* defines a high-rise as "a building having many storeys".
- The *International Conference on Fire Safety in High-Rise Buildings* defined a high-rise as "any structure where the height can have a serious impact on evacuation"
- In the U.S., the National Fire Protection Association defines a high-rise as being higher than 75 feet (23 metres), or about 7 stories.
- Most building engineers, inspectors, architects and similar professions define a high-rise as a building that is at least 75 feet (23 m) tall.

### History

High-rise apartment buildings had already appeared in antiquity: the *insulae* in ancient Rome and several other cities in the Roman Empire, some of which might have reached up to ten or more storeys, one reportedly having 200 stairs. Because of the destruction caused by poorly built high-rise *insulae* collapsing, several Roman emperors, beginning with Augustus (r. 30 BC – 14 AD), set limits of 20–25 metres for multi-storey buildings, but met with limited success, as these limits were often ignored despite the likelihood of taller *insulae* collapsing. The lower floors were typically occupied by either shops or wealthy families, while the upper storeys were rented out to the lower classes. Surviving Oxyrhynchus Papyri indicate that seven-storey buildings even existed in provincial towns, such as in third century AD Hermopolis in Roman Egypt.

In Arab Egypt, the initial capital city was Fustat. It housed many high-rise residential buildings, some seven stories tall that could reportedly accommodate hundreds of people. Al-Muqaddasi, in the 10th century, described them as resembling minarets, while NasirKhusraw, in the early 11th century, described some of them rising up to 14 storeys, with roof gardens on the top storey complete with ox-drawn water wheels for irrigating them. By the 16th century, Cairo also had high-rise apartment buildings where the two lower floors were for commercial and storage purposes and the multiple stories above them were rented out to tenants.

The skyline of many important medieval cities was dominated by large numbers of high-rising urban towers, which fulfilled defencive but also representative purposes. The residential Towers of Bologna numbered between 80 to 100 at a time, the largest of which still rise to 97.2 m. In Florence, a law of 1251 decreed that all urban buildings should be reduced to a height of less than 26 m, the regulation immediately put into effect. Even medium-sized towns such as San Gimignano are known to have featured 72 towers up to 51 m in height.

Tower blocks were built in the Yemeni city of Shibam in the 16th century. The houses of Shibam are all made out of mud bricks, but about five hundred of them are tower houses, which rise five to sixteen stories high, with each floor having one or two apartments. This technique of building was implemented to protect residents from Bedouin attacks. While Shibam has existed for around two thousand years, most of the city's houses date from the 16th century. The city has the tallest mud buildings in the world, some more than 30 metres (100 feet) high. Shibam has been called "one of the oldest and best examples of urban planning based on the principle of vertical construction" or "Manhattan of the desert".

Currently, the tallest high-rise apartment building in the world is Chicago's John Hancock Centre, constructed under the supervision of Skidmore, Owings & Merrill and completed in 1969. The building has 100 stories and stands 344 metres tall.

## Modern Development

***United Kingdom:*** Tower blocks were built in the UK after the Second World War. The first residential tower block, "The Lawn" was constructed in Harlow, Essex in 1951; it is now a Grade II listed building. In many cases tower blocks were seen as a "quick-fix" to cure problems caused by crumbling and unsanitary 19th-century dwellings or to replace buildings destroyed by German aerial bombing. Initially, they were welcomed, and their excellent views made them popular living places. Later, as the buildings themselves deteriorated, they grew a reputation for being undesirable low cost housing, and many tower blocks saw rising crime levels, increasing their unpopularity. One response to this was the great increase in the number of housing estates built, which in turn brings its own problems. In the UK, tower blocks particularly lost popularity after the partial collapse of Ronan Point in 1968. Glasgow, the largest city in Scotland, is believed to contain the highest concentration of tower blocks in the UK: examples include the Hutchesontown C blocks in the Gorbals, the 20-storey blocks in Sighthill, and the 31-storey Red Road (flats) in the city's north east. However, on the whole, London has the largest number of high-rise residential buildings in the UK.

### *The Post-War British Tower Block Vision*

Post-war Britain was the stage for a tower block "building boom"; from the 1950s to the late 1970s there was a dramatic increase in tower block construction. During this time, local authorities desired to impress their voters by building futuristic and imposing tower blocks, which would signify post-war progress. Both Patrick Dunleavy and Lynsey Hanley agree that architects and planners were influenced by Le Corbusier's promotion of high-rise architecture. The modern tower blocks were to include features that would foster desired forms of resident interaction, an example being the inclusion of Le Corbusier's streets in the sky in some estates.

As well as inspiring residents, local authority planners believed that the way tower blocks were constructed would save money. Generally, the tower blocks were built on cheap greenfield land skirting established cities. Although the property prices for these periphery sites were markedly cheaper than their inner city counterparts, they often had little access to public amenities, such as public transport. It was thought that the implementation of industrialised building techniques would lower costs too, as similar tower blocks would be replicated over many sites. Uniform and standardised parts, such as toilet fittings and door handles, would be fitted throughout many tower blocks; planners deemed that buying in bulk would reduce overall costs.

Another key aspect of the tower block vision was the Brutalist architectural method, popular with architects and planners at the time. The Brutalist emphasis led to the construction of stark and striking tower blocks with large sections of exposed concrete. Concrete was to be an integral part of the tower block designs; it could be poured on site, offering boundless flexibility to the building designers. To the planners, concrete was a silver bullet for the construction process – it was economical, and "was vaunted as being long-lasting, if not indestructible".

### The Post War British Tower Block Reality

Coleman's 1985 work argues that in trying to emulate Le Corbusier's ideas, the tower block planners only succeeded in encouraging social problems. Although architects and local authorities intended the opposite, tower blocks quickly became, as Hanley sharply stated, 'slums in the sky'. Due to demanding deadlines, complicated construction practices were rushed and many tower blocks experienced structural decay as a result – roofs leaked, concrete suffered spalling, steel corroded, and damp penetrated the buildings. Unfortunately, by replicating tower blocks across the nation, planners 'disastrously' replicated design faults. In many tower blocks, concrete quickly exhibited signs of decay; cracks soon formed and destabilised the buildings. The partial collapse of the Ronan Point tower block is an infamous example of the hasty and substandard construction that occurred in a number of the towers. The tower blocks quickly lost their "futuristic" look; concrete turned from the crisp white the designers had imagined to a dull grey, stained by pollution.

Poor design decisions ruined the anticipated benefits of the buildings. Open spaces, which were supposed to benefit the residents, were instead unattractive, unused and inadequately supervised. Residents felt it was difficult to maintain the large open spaces around the blocks because they realistically belonged to no one. Social problems increased as the tower blocks quickly degraded because of poor maintenance and an insecure communal environment. Apart from frequent break-downs, communal lifts were a source of fear for people travelling alone. It was a rarity to "enter a clean-smelling, undefaced lift". The tower blocks, many of which were on the periphery of the city, made residents feel isolated and cut off from society. Outsiders and newcomers were also affected; they felt the overbearing design of the tower blocks made them fearsome and unsociable.

Power argues that as a direct consequence of their design and construction, security problems were prevalent in many of the tower blocks. Break-ins, vandalism and muggings were common, which were aided by the buildings' concealed areas, the mazes of internal corridors, and dark corners. Police were often required in the tower blocks, but their infrequent presence did little to pacify towers rife with delinquency. To contain disruptive behaviour, local authorities began to place "problem families" in the same blocks; Hanley argues that this policy only led to "further alienation ... nihilism and a creeping sense of lawlessness". Dunleavy seconds this, suggesting that the mental health of long-term tower block residents may have been detrimentally affected.

While local authorities and their architects intended to create tower blocks that encouraged harmonious and vibrant communities, often the results were far from ideal. Post-war tower blocks were compromised from the outset by a combination of faults: local authorities advocated impractical architectural methods; design and construction faults were frequently reproduced; and there appeared to be a lack of understanding about the social consequences of certain design features. Collectively, these oversights transformed many tower blocks into undesirable places to live.

### *Fire Safety*

The lives of residents in London Tower blocks are at risk because London councils have not done proper fire safety checks and the problem is probably widespread in the UK. North Ayrshire Council has decided to remove plastic cladding from its buildings following a fatal fire in a tower block. Poor maintenance of electrical equipment, fire doors and other features has caused widespread fire hazards and residents are asked to pay part of the cost of putting this right even though the problem is due to poor maintenance and errors in design.

Tower blocks may be inherently more prone to casualties from a fire because people living on higher floors cannot escape fires easily and the fire brigade cannot reach the higher floors quickly. In buildings with more than a hundred residents, ensuring that every single resident acts responsibly to minimize fire risk is impossible; poorer residents in tower blocks may be tempted to use cheaper flammable fuels rather than electricity, they are also more likely to be smokers (carelessness with cigarettes is a major cause of home fires), and they are more likely to have old furniture, not made to modern fire safety standards.

Fire safety legislation introduced in London in 2006 requires new high rise buildings to be built to higher safety standards with sprinkler systems; these new buildings are mainly luxury accommodation for richer people. The legislation is not retroactive and does not protect poor people living in existing tower blocks.

## *Towards the Present Day*

***Figure:*** *The iconic 31-storey Red Road estate in Glasgow, the tallest public housing towers in Europe at the time of their construction, like many 1960s British high-rises, are now condemned.*

In recent years, some council or ex-council high-rises in the United Kingdom, including Trellick Tower, Keeling House, Sivill House and The Barbican Estate, have become popular with young professionals due to their excellent views, desirable locations and architectural pedigrees, and now command high prices. There are plans to redevelop the Little London and Lovell Park areas on the fringes of Leeds city centre into luxury flats for 'Young Urban Professionals'.

The plans entail demolishing all of the council housing and refurbishing the highrise flats. This demand has led to many councils rethinking plans regarding their demolition. In Glasgow, similar initiatives have taken place – in 2011, the Glasgow Housing Association saved one of a cluster of three condemned tower blocks in the Ibrox area with the help of government grants to transform it into desirable rented housing for young professionals. Glasgow itself has taken a more measured approach to its high-rise housing stock, eschewing the mass demolition programmes being practiced in other British cities – favouring selective refurbishment of estates with secure futures and on buildings in good structural condition, only razing towers where absolutely necessary.

After a gap of around 30 years, new high-rise flats are once again being built in Belfast, Birmingham (some are for wealthy people, e.g. The Mailbox, others are student accommodation), Cardiff, Aberdeen, Glasgow, Leeds, Liverpool, London, Manchester, and Newcastle upon Tyne; but this time sometimes for wealthy professionals, rather than the "lower classes." Their developers market these properties by using the American term 'apartment buildings', perhaps in an effort to distance these newer buildings from the older tower blocks from the 1950s and 1960s. These are usually taller than their older counterparts and generally built in and around these provincial city centres. They are often glass and aluminium clad. Tonight with Trevor McDonald highlighted that in Leeds and Manchester (perhaps the cities that had seen most development) only approximately half were occupied and with owner occupation often being as low as 10%. In Southend there are controversial plans to build new tower blocks for social housing, a supporter claims this will be cheaper than paying to house poor people with private landlords while opponents fear the creation of Ghettos.

Tower blocks in Northern Ireland were never built to the frequency as they were in other cities in Britain and Ireland. Most tower blocks and flat complexes are found in Belfast and Derry, although many of these have been demolished in recent years and replaced with traditional public housing units. The Divis flats complex in west Belfast was built in between 1968 and 1972 was demolished in the early 1990s as the residents demanded new houses due to mounting problems with the flats. Divis Tower, built separately in 1966, still stands, however; and, in 2007, work began to convert the former British Army base at the top two floors into new dwellings.

In the north of the City, the iconic 7 towers complex in the New Lodge remains, although so too the problems that residents face, such as poor piping and inadequate sanitation. Farther north, the 4 tower blocks in Rathcoole still dominate the local skyline, while in south Belfast, the tower blocks in Seymour Hill also remain standing.

### *Ireland*

Apart from a limited number of blocks in Londonderry and Larne all of the post-war high rise dwellings in Northern Ireland (UK) were situated in Belfast. Most of that city's tower blocks were built by the Northern Ireland Housing Trust (NIHT) as part of overspill housing schemes, the first such development being the pair of point blocks in East Belfast's Cregagh estate. These eleven storey towers were completed in 1961 and were the first tall council housing blocks on the island of Ireland. The NIHT also designed the inner city Divis Flats complex. The twenty storey high Divis Tower was for several decades Ireland's tallest residential building, having since being surpassed by the privately owned Obel Tower in Belfast city centre. The six to eight storey deck access flats that comprised most of the Divisestate were of poor build quality and were all demolished by the early 1990s. Similar slab blocks were built by the NIHT in East Belfast (Tullycarnet) and Londonderry's Bogside area, all four of which were eventually demolished.

Belfast Corporation constructed seven tower blocks on the former Victoria Barracks site in the New Lodge district. While the Corporation built some medium rise flat complexes as part of slum clearance schemes New Lodge would be their only high rise project in the inner city. They constructed three additional tower blocks in outlying areas of the city during the 1960s, two being in Mount Vernon in North Belfast and one being in the Clarawood estate, East Belfast.

High rise accommodation was also built for hospital workers at Belfast's Royal Hospital, consisting of three thirteen storey towers. Of the three point blocks developed by Larne Borough Council in the late 1960s, only one of the sixteen storey buildings remains.

In Dublin, the Ballymun Flats were built between 1966 and 1969 consisting of seven 15-storey towers, nineteen 8-storey blocks and ten 4-storey blocks. These were the "seven towers" referred to in the U2 song "Running to Stand Still". Inner Dublin flat complexes include Sheriff Street (demolished), Fatima Mansions (demolished and redeveloped), St. Joseph's Gardens (demolished; replaced by Killarney Court flat complex), St. Teresa's Gardens, Dolphin House, Liberty House, St. Michael's Estate and O'Devaney Gardens and a lot more mainly throughout the North and South Inner City of Dublin. Suburban flat complexes were built exclusively on the northside of the city in Ballymun, Coolock and Kilbarrack. These flats were badly affected by a heroin epidemic that hit working-class areas of Dublin in the 1980s and early 90s.

Over the last five years the largest cities such as Dublin, Cork, Limerick and Galway have witnessed new large apartment building. Some large towns such as Navan, Drogheda, Dundalk and Mullingar have also witnessed lots of modern apartments being built.

### Eastern Europe

Russia is currently undergoing a dramatic buildout, growing a commercially shaped skyline.

### East Asia

Residential tower complexes are common in Asian countries such as China, Singapore, Japan, and Korea as urban densities are very high. In Singapore and urban Hong Kong, land prices are so high that almost the entire population lives in high-rise apartments. In fact, over 60% of Hong Kong residents live in apartments, many of them condominiums.

Sarah Williams Goldhagen (2012) celebrated the work of innovative architecture firms such as WOHA (based in Singapore), Mass Studies (based in Seoul), Amateur Architecture Studio (based in Hangzhou), and the New York City-based Steven Holl in the transformation of residential towers into "vertical communities" or "vertical cities in the sky" providing aesthetic, unusually designed silhouettes on the skyline, comfortable private spaces and attractive public spaces. None of these "functional, handsome, and humane high-rise residential buildings" are affordable housing.

### South Korea

In South Korea the tower blocks are called "Apartment Complex *(Apartment Danji)*". The first residential towers began to be built after the Korean War. The South Korean government needed to build many apartment complexes in the cities to be able to accommodate the citizens. In the 60 years since, as the population increased considerably, tower blocks have become more common. This time however the new tower blocks integrated shopping malls, parking system and other convenient facilities.

Tower Palace in Seoul, South Korea, is the tallest apartment complex in Asia.

In Seoul (Cho 2008), approximately 80% of its residents live in apartment complexes which comprise 98% of recent residential construction. Seoul proper is noted for its population density, almost twice that of New York and eight times greater than Rome, though slightly less than Paris. Its metropolitan area is the densest in the OECD.

### China

The 2012 Pritzker Prize was awarded to Chinese architect Wang Shu. Among his winning designs is the Vertical Courtyard Apartments, six 26-storey towers built in Hangzhou, China, by his architectural firm, Amateur Architecture Studio. "These towers were designed to house two-storey apartments, in which every inhabitant would enjoy "the illusion of living on the second floor", accomplished by folding concrete floor planes (like "bamboo mats," claims the firm), so that every third storey opens into a private courtyard. In the larger towers, the two-storey units are stacked slightly askew, adding to the visual interest of the variegated façades (Goldhagen 2012)."

### Canada

In Canada large multi-family buildings are usually known as *apartment buildings* or *apartment blocks* if they are rented from one common landowner, or *condominiums* or *condo towers* if each dwelling unit is individually owned; they may be called *low-rise* (or *walk-up*), *mid-rise, high-rise,* or *skyscraper* depending on their height. Tall residential towers are a staple building type in all large cities. Their relative prominence in Canadian cities varies substantially, however. In general, more populated cities have more high-rises than smaller cities, due to a relative the scarcity of land and a greater demand for housing.

However, some cities such as Quebec City, Ottawa, and Halifax have fewer high-rise buildings due to several factors: a focus on historic preservation, height restrictions, and lower growth rates. In middle-sized cities with a relatively low population density, such as Calgary, Edmonton, Winnipeg, or Hamilton, there are more apartment towers but they are greatly outnumbered by single-family houses. Most of the largest residential towers in Canada are found in Toronto, Montreal, and Vancouver—the country's most densely-populated cities.

Toronto contains the second largest concentration of high-rise apartment buildings in North America (after New York). In Canada, like in other New World countries, but unlike Western Europe, most high-rise towers are located in the city centre (or "downtown"), where smaller, older buildings were demolished to make way in redevelopment schemes. Toronto, however has many high-rise apartment and condominium communities in its denser suburbs in addition to downtown. Greenfield high-rise development in the suburbs of Canadian cities is much more rare, although it does happen, particularly in transit-oriented developments.

### United States

In the United States tower blocks are commonly referred to as "midrise" or "highrise apartment buildings", depending on their height, while buildings that house fewer flats (apartments), or are not as tall as the tower blocks, are called "lowrise apartment buildings".

Some of the first residential towers were the Castle Village towers in New York City completed in 1939. Their cross-shaped design was copied in towers in Parkchester and Stuyvesant Town residential developments.

The government's experiments in the 1960s and 70s to use high-rise apartments as a means of providing the housing solution for the poor resulted in a spectacular failure. All but a few high-rise housing projects in the nation's largest cities, such as Cabrini–Green and Robert Taylor Homes in Chicago, Penn South in New York City and the Desire projects in New Orleans, fell victim to the "ghettofication" and are now being torn down, renovated, or replaced.

In contrast to their public housing cousins, commercially developed high-rise apartment buildings continue to flourish in cities around the country largely due to high land prices and the housing boom of the 2000s. The Upper East Side in New York City and Chicago's Gold Coast, both featuring high-rise apartments, are the wealthiest urban neighbourhoods in the United States. Currently, the tallest high-rise apartment building and tower block in the world is Chicago's John Hancock Centre, constructed by Bangladeshi engineer Fazlur Khan in 1969. The building has 100 storeys and stands at 344 metres tall.

### Australia

High-rise living in Australia was limited to small pockets of bohemian inner Sydney until the 1960s, where a short-lived fashion saw public housing tenants located in new high-rise developments, especially in Sydney and Melbourne. The buildings pictured along with four other 16-storey blocks were constructed on behalf of the Royal Australian Navy and were available to sailors and their families for accommodation. Due to social problems within these blocks the Navy left and the Department of Housing took charge and flats were let to low income and immigrant families. During the 1980s many people escaping communism in Eastern block countries were housed in these buildings. Developers have enthusiastically adopted the term "apartment" for these new high-rise blocks, perhaps to avoid the stigma still attached to housing commission flats.

### Streets in the Sky

"Streets in the sky" redirects here. For the third studio album by UK rock band The Enemy.

Streets in the sky is a style of architecture that emerged in Britain in the 1960s and 1970s. Generally built to replace run-down terraced housing, the new designs included not only modern improvements such as inside toilets, but also shops and other community facilities within high-rise blocks. Examples of the buildings and developments are Trellick Tower, Balfron Tower, Robin Hood Gardens and Keeling House in London, Hunslet Grange in Leeds and Park Hill, Sheffield. These were an attempt to develop a new architecture, differentiated from earlier large housing estates, such as Quarry Hill flats in Leeds. Alison and Peter Smithson were the architects of Robin Hood Gardens. Another large example, the Aylesbury Estate in South London, built in 1970, is about to be demolished. The Hulme Crescents in Manchester were the largest social housing scheme in Europe when built in 1972 but lasted just 22 years. The Crescents are recognised as one of the worst social housing schemes in British history, marred by design and practical problems; they severely challenged the viability of the Streets in the Sky ideal.

### *Deck Access*

Deck access is a type of flat that is accessed from a walkway that is open to the elements, as opposed to flats that are accessed from fully enclosed internal corridors. Deck access blocks of flats are usually fairly low-rise structures. The decks can vary from simple walkways, which may be covered or uncovered, to decks wide enough for small vehicles. The best known example of deck-access flats in the UK is Park Hill, Sheffield, where the decks are wide enough to allow electric vehicles; however, the design is inspired by French Modernist architect Le Corbusier, particularly his Unite D'Habitation in Marseilles.

### *Green Tower Blocks*

Green tower blocks have some scheme of living plants or green roofs or solar panels on their roofs and/or incorporate other environmentally friendly design features.

## Building Code

A building code, or building control, is a set of rules that specify the minimum acceptable level of safety for constructed objects such as buildings and nonbuilding structures. The main purpose of building codes are to protect public health, safety and general welfare as they relate to the construction and occupancy of buildings and structures. The building code becomes law of a particular jurisdiction when formally enacted by the appropriate governmental or private authority.

Building codes are generally intended to be applied by architects, engineers, constructors and regulators but are also used for various purposes by safety inspectors, environmental scientists, real estate developers, subcontractors, manufacturers of building products and materials, insurance companies, facility managers, tenants, and others. Codes regulating the design and construction of structures where adopted into law. Codes in developed western nations can be quite complex and exhaustive. They began in ancient times and have been developing ever since. In the USA the main codes are the International Commercial or Residential Code [ICC/IRC], electrical codes and plumbing, mechanical codes. Other codes may include fire, health, transportation, manufacturing, and other regulations/regulators/testers such as UL; Underwriters Labs. In essence they are minimum standards of design and implementation. Designers use ICC/IRC standards out of substantial reference books during design. Building departments review plans submitted to them before construction, issue permits [or not] and inspectors verify compliance to these standards at the site during construction.

There are often additional codes or sections of the same building code that have more specific requirements that apply to dwellings or places of business and special construction objects such as canopies, signs, pedestrian walkways, parking lots, and radio and television antennas.

### *Types of Building Codes*

The practice of developing, approving, and enforcing building codes varies considerably among nations. In some countries building codes are developed by the government agencies or quasi-governmental standards organisations and then enforced across the country by the central government. Such codes are known as the national building codes (in a sense they enjoy a mandatory nation-wide application).

In other countries, where the power of regulating construction and fire safety is vested in local authorities, a system of model building codes is used. Model building codes have no legal status unless adopted or adapted by an authority having jurisdiction. The developers of model codes urge public authorities to reference model codes in their laws, ordinances, regulations, and administrative orders. When referenced in any of these legal instruments, a particular model code becomes law. This practice is known as *adoption by reference.* When an adopting authority decides to delete, add, or revise any portions of the model code adopted, it is usually required by the model code developer to follow a formal adoption procedure in which those modifications can be documented for legal purposes.

***Figure:*** *"Sutyagin's skyscraper" (Íåáîñêð¸áÑóòÿãèíà) - supposedly world's tallest wooden single-family house - found to be in violation of fire codes by the city of Arkhangelsk, Russia, and was demolished.[1]*

There are instances when some local jurisdictions choose to develop their own building codes. At some point in time all major cities in the United States had their own building codes. However due to ever increasing complexity and cost of developing building regulations, virtually all municipalities in the country have chosen to adopt model codes instead. For example, in 2008 New York City abandoned its proprietary *1968 New York City Building Code* in favour of a customized version of the International Building Code. The City of Chicago remains the only municipality in America that continues to use a building code the city developed on its own as part of the *Municipal Code of Chicago*.

In Europe, the Eurocode is a pan-European building code that has superseded the older national building codes. Each country now has "country annexes" to localize the contents of the Eurocode. Similarly, in India, each municipality and urban development authority has its own building code, which is mandatory for all construction within their jurisdiction. All these local building codes are variants of a National Building Code, which serves as model code proving guidelines for regulating building construction activity.

### Scope

Building codes generally include:

- Standards for structure, placement, size, usage, wall assemblies, fenestration size/locations, egress rules, size/location of rooms, foundations, floor assemblies, roof structures/assemblies, energy efficiency, stairs and halls, mechanical, electrical, plumbing, site drainage & storage, appliance, lighting, fixtures standards, occupancy rules, even swimming pool regulations. Sounds like a lot but there are nearly infinite ways to build it in a hazardous fashion; falls, electrocution, contamination, explosion, fire.
- Rules regarding parking and traffic impact
- Fire code rules to minimize the risk of a fire and to ensure safe evacuation in the event of such an emergency
- Requirements for earthquake, hurricane, flood, and tsunami resistance, especially in disaster prone areas or for very large buildings where a failure would be catastrophic
- Requirements for specific building uses (for example, storage of flammable substances, or housing a large number of people)
- Energy provisions and consumption
- Grandfathering provisions: Unless the building is being renovated, the building code usually does not apply to existing buildings.
- Specifications on components
- Allowable installation methodologies
- Minimum and maximum room and exit sizes and location
- Qualification of individuals or corporations doing the work
- For high structures, anti-collision markers for the benefit of aircraft

Building codes are generally separate from zoning ordinances, but exterior restrictions (such as setbacks) may fall into either category.

### Prescriptive vs. Performance

These requirements are usually a combination of prescriptive requirements that spell out exactly how something is to be done, and performance requirements which just outline what the required level of performance is and leave it up to the designer how this is achieved. Historically they are very reactive in that when a problem occurs the building codes change to ensure that the problem never happens again. In recent years there has been a move amongst many building codes to move to more performance requirements and less prescriptive requirements.

Traditionally building codes were generally short non complex interrelated sets of rules. They generally included reference to hundreds of other codes, standards and guidelines that specify the details of the component or system design, specify testing requirements for components, or outline good engineering practice. These detailed codes required a great deal of specialisation to interpret, and also greatly constrained change and innovation in building design. In recent years several countries, beginning with Australia, have moved to much shorter objective based buildings codes. Rather than prescribing specific details, objective codes lists a series of objectives all buildings must meet while leaving open how these objectives will be met. When applying for a building permit the designers must demonstrate how they meet each objective.

### History

***Antiquity:*** Building codes have a long history. What is generally accepted as the first building code was in the Code of Hammurabi which specified:

- 229. If a builder builds a house for someone, and does not construct it properly, and the house which he built falls in and kills its owner, then that builder shall be put to death.
- 230. If it kills the son of the owner, the son of that builder shall be put to death.
- 231. If it kills a slave of the owner, then he shall pay, slave for slave, to the owner of the house.
- 232. If it ruins goods, he shall make compensation for all that has been ruined, and inasmuch as he did not construct properly this house which he built and it fell, he shall re-erect the house from his own means.
- 233. If a builder builds a house for someone, even though he has not yet completed it; if then the walls seem toppling, the builder must make the walls solid from his own means.

The Law of Moses stipulated a specific construction requirement which is also an early form of a building code. The Bible book of Deuteronomy, chapter 22 verse 8, states:

- "In case you build a new house, you must also make a parapet for your roof, that you may not place bloodguilt upon your house because someone falling might fall from it."

### Colonial Era

- Rebuilding of London Acts 1666&1670
- 16th-century Spanish Laws of the Indies

### Nineteenth-Century Building Laws

The great changes in societies in Europe, the Industrial Revolution, the end of slavery in the United States, and immigration to the United States brought about the enactment of a number of building laws during the nineteenth century.

Baltimore passed its first building code in 1859. The Great Baltimore Fire occurred in February, 1904. Subsequent changes were made that matched other cities. In 1904, a Handbook of the Baltimore City Building Laws was published. It served as the building code for four years. Very soon, a formal building code was drafted and eventually adopted in 1908.

### France

In Paris, great blocks of apartments were erected under the Second Empire (1852–70). The height of buildings was limited by law, so they were usually five or six stories at most.

### Germany and Austria

Germany and Austria generally followed the French plan.

### United Kingdom

The most important statutes of this kind in the United Kingdom were the London Building Act of 1844 and the Public Health Act of 1875. The Metropolitan Buildings Office was established in 1845 to enforce building regulations.

### United States

The major model building codes used in the United States are developed by the International Code Council (ICC), which have 14 sets of *International codes*, or *i-codes*, including the International Building Code (IBC), the International Residential Code, the International Fire Code, the International Energy Conservation Code, the International Plumbing Code, the International Mechanical Code and others.

## Building Material

Building material is any material which is used for construction purposes. Many naturally occurring substances, such as clay, rocks, sand, and wood, even twigs and leaves, have been used to construct buildings. Apart from naturally occurring materials, many man-made products are in use, some more and some less synthetic.

The manufacture of building materials is an established industry in many countries and the use of these materials is typically segmented into specific speciality trades, such as carpentry, insulation, plumbing, and roofing work. They provide the make-up of habitats and structures including homes.

### *Naturally Occurring Substances*

***Brush:*** Brush structures are built entirely from plant parts and are generally found in tropical and sub-tropical areas, such as rainforests, where very large leaves can be used in the building. These are built mostly with branches, twigs and leaves, and bark, similar to a beaver's lodge. These were variously named wikiups, lean-tos, and so forth.

An extension on the brush building idea is the wattle and daub process in which clay soils or dung, usually cow, are used to fill in and cover a woven brush structure. This gives the structure more thermal mass and strength. Wattle and daub is one of the oldest building techniques. Many older timber frame buildings incorporate wattle and daub as non load bearing walls between the timber frames.

### *Ice*

Ice was used by the Inuit for igloos, but has also been used for ice hotels as a tourist attraction in northern areas that might not otherwise see many winter tourists.

### *Mud and Clay*

Clay based buildings usually come in two distinct types. One being solid monolithic structural walls such as Rammed earth, sod, and cob. The other being walls built by stacking building blocks or bricks by forming the clay mixture into individual units which are then dried or fired, examples include compressed earth blocks and masonry brickwork.

Other uses of clay in building is combined with straws to create light clay insulation and plasters.

### Monolithic Clay Structure

The amount of and type of each material in the mixture used leads to different styles of buildings. The deciding factor is usually connected with the quality of the soil being used. Larger amounts of clay are usually employed in the *cob/adobe* style, while low clay soil is usually associated with *sod* building. The other main ingredients include more or less sand/gravel and straw/grasses. *Rammed earth* is both an old and newer take on creating walls, once made by compacting clay soils between planks by hand; nowadays forms and mechanical pneumatic compressors are used.

Soil, and especially clay, provides good thermal mass; it is very good at keeping temperatures at a constant level. Homes built with earth tend to be naturally cool in the summer heat and warm in cold weather. Clay holds heat or cold, releasing it over a period of time like stone. Earthen walls change temperature slowly, so artificially raising or lowering the temperature can use more resources than in say a wood built house, but the heat/coolness stays longer.

Peoples building with mostly dirt and clay, such as cob, sod, and adobe, created homes that have been built for centuries in western and northern Europe, Asia, as well as the rest of the world, and continue to be built, though on a smaller scale. Some of these buildings have remained habitable for hundreds of years.

### Structural Clay Blocks

Bricks or *compressed earth blocks* are used for building more frequently in industrialized society since the building blocks can be manufactured off site in a centralized location at a brickworks and transported to multiple building locations. These blocks can also be monetized more easily and sold.

Structural bricks are almost always made using clay, often clay soil is the only ingredient used, but other ingredients can include sand, lime, concrete, stone and other binders. The formed or compressed block is then air dried, fired, or kiln dried. Kiln fired clay bricks are considered a ceramic material. Clay bricks can be solid or have hollow cavities to aid in drying and make them lighter and easier to transport.

The individual bricks are placed upon each other in rows using mortar, grout, and clay slips. Successive rows being used to build up walls, arches, and other structures.

Brick walls can be built substantially thinner than cob/adobe while keeping the same vertical strength. They require more energy to create but are easier to transport and store, and are lighter than stone blocks. Romans were fond of building with brick. Building with brick gained much popularity in the mid-18th century and 19th centuries. This was probably due to the fact that it was much more flame retardant than wood in the ever crowding cities, and fairly cheap to produce. Another type of block replaced clay bricks in the late 20th century. It was the Cinder block. Made mostly with concrete.

### Sand

Sand is used with cement, and sometimes lime, to make mortar for masonry work and plaster. Sand is also used as a part of the concrete mix. An important low-cost building material in countries with high sand content soils is the Sandcrete block, which is weaker but cheaper than fired clay bricks.

### Stone or Rock

Rock structures have existed for as long as history can recall. It is the longest lasting building material available, and is usually readily available. There are many types of rock throughout the world, all with differing attributes that make them better or worse for particular uses. Rock is a very dense material so it gives a lot of protection too; its main drawback as a material is its weight and awkwardness. Its energy density is also considered a big drawback, as stone is hard to keep warm without using large amounts of heating resources. Dry-stone walls have been built for as long as humans have put one stone on top of another. Eventually, different forms of mortar were used to hold the stones together, cement being the most commonplace now.

The granite-strewn uplands of Dartmoor National Park, United Kingdom, for example, provided ample resources for early settlers. Circular huts were constructed from loose granite rocks throughout the Neolithic and early Bronze Age, and the remains of an estimated 5,000 can still be seen today. Granite continued to be used throughout the Medieval period and into modern times. Slate is another stone type, commonly used as roofing material in the United Kingdom and other parts of the world where it is found.

Stone buildings can be seen in most major cities; some civilizations built entirely with stone such as the Egyptian and Aztec pyramids and the remains of the Inca civilization.

### Thatch

Thatch is one of the oldest of building materials known; grass is a good insulator and easily harvested. Many African tribes have lived in homes made completely of grasses and sand year-round. In Europe, thatch roofs on homes were once prevalent but the material fell out of favour as industrialization and improved transport increased the availability of other materials. Today, though, the practice is undergoing a revival. In the Netherlands, for instance, many new buildings have thatched roofs with special ridge tiles on top.

### Wood/Timber

Wood has been used as a building material for thousands of years in its natural state. Today, engineered wood is becoming very common in industrialized countries.

Wood is a product of trees, and sometimes other fiberous plants, used for construction purposes when cut or pressed into lumber and timber, such as boards, planks and similar materials. It is a generic building material and is used in building just about any type of structure in most climates. Wood can be very flexible under loads, keeping strength while bending, and is incredibly strong when compressed vertically. There are many differing qualities to the different types of wood, even among same tree species. This means specific species are better suited for various uses than others. And growing conditions are important for deciding quality.

"Timber" is the term used for construction purposes except the term "lumber" is used in the United States. Raw wood (a log, trunk, bole) becomes timber when the wood has been "converted" (sawn, hewn, split) in the forms of minimally-processed logs stacked on top of each other, timber frame construction, and light-frame construction. The main problems with timber structures are fire risk and moisture-related problems.

In modern times softwood is used as a lower-value bulk material, whereas hardwood is usually used for finishings and furniture. Historically timber frame structures were built with oak in western Europe, recently douglas fir has become the most popular wood for most types of structural building.

Many families or communities, in rural areas, have a personal woodlot from which the family or community will grow and harvest trees to build with or sell. These lots are tended to like a garden. This was much more prevalent in pre-industrial times, when laws existed as to the amount of wood one could cut at any one time to ensure there would be a supply of timber for the future, but is still a viable form of agriculture.

## Man-Made Substances

***Cement Composites:*** Cement bonded composites are made of hydrated cement paste that binds wood, particles, or fibres to make pre-cast building components. Various fiberous materials, including paper, fibreglass, and carbon-fibre have been used as binders.

Wood and natural fibres are composed of various soluble organic compounds like carbohydrates, glycosides and phenolics. These compounds are known to retard cement setting. Therefore, before using a wood in making cement bonded composites, its compatibility with cement is assessed.

Wood-cement compatibility is the ratio of a parameter related to the property of a wood-cement composite to that of a neat cement paste. The compatibility is often expressed as a percentage value. To determine wood-cement compatibility, methods based on different properties are used, such as, hydration characteristics, strength, interfacial bond and morphology. Various methods are used by researchers such as the measurement of hydration characteristics of a cement-aggregate mix; the comparison of the mechanical properties of cement-aggregate mixes and the visual assessment of microstructural properties of the wood-cement mixes. It has been found that the hydration test by measuring the change in hydration temperature with time is the most convenient method. Recently, Karade et al.have reviewed these methods of compatibility assessment and suggested a method based on the 'maturity concept' i.e. taking in consideration both time and temperature of cement hydration reaction.

## Concrete

Concrete is a composite building material made from the combination of aggregate and a binder such as cement. The most common form of concrete is Portland cement concrete, which consists of mineral aggregate (generally gravel and sand), portland cement and water.

After mixing, the cement hydrates and eventually hardens into a stone-like material. When used in the generic sense, this is the material referred to by the term "concrete".

For a concrete construction of any size, as concrete has a rather low tensile strength, it is generally strengthened using steel rods or bars (known as rebars). This strengthened concrete is then referred to as reinforced concrete. In order to minimise any air bubbles, that would weaken the structure, a vibrator is used to eliminate any air that has been entrained when the liquid concrete mix is poured around the ironwork. Concrete has been the predominant building material in the modern age due to its longevity, formability, and ease of transport. Recent advancements, such as insulating concrete forms, combine the concrete forming and other construction steps (installation of insulation). All materials must be taken in required proportions as described in standards.

### Fabric

The tent is the home of choice among nomadic groups all over the world. Two well-known types include the conical teepee and the circular yurt. The tent has been revived as a major construction technique with the development of tensile architecture and synthetic fabrics. Modern buildings can be made of flexible material such as fabric membranes, and supported by a system of steel cables, rigid or internal, or by air pressure.

### Foam

***Figure:*** *Foamed plastic sheet to be used as backing for firestop mortar at CIBC bank in Toronto.*

Recently, synthetic polystyrene or polyurethane foam has been used in combination with structural materials, such as concrete. It is lightweight, easily shaped, and an excellent insulator.

Foam is usually used as part of a structural insulated panel, wherein the foam is sandwiched between wood or cement or insulating concrete forms.

### Glass

Glassmaking is considered an art form as well as an industrial process or material. Clear windows have been used since the invention of glass to cover small openings in a building. Glass panes provided humans with the ability to both let light into rooms while at the same time keeping inclement weather outside.

Glass is generally made from mixtures of sand and silicates, in a very hot fire stove called a kiln, and is very brittle.

Additives are often included the mixture used to produce glass with shades of colours or various characteristics (such as bulletproof glass or light emittance).

The use of glass in architectural buildings has become very popular in the modern culture. Glass "curtain walls" can be used to cover the entire facade of a building, or it can be used to span over a wide roof structure in a "space frame".

These uses though require some sort of frame to hold sections of glass together, as glass by itself is too brittle and would require an overly large kiln to be used to span such large areas by itself.

### Gypcrete

Gypcrete is a mixture of gypsum plaster and fibreglass rovings. Although plaster and fibres fiborous plaster have been used for many years, especially for ceilings, it was not until the early 1990s that serious studies of the strength and qualities of a walling system Rapidwall, using a mixture of gypsum plaster and 300mm plus fibreglass rovings, were investigated. It was discovered, through testing at the University of Adelaide, that these walls had significant, load bearing, shear and lateral resistance together with earthquake-resistance, fire-resistance, and thermal properties.

With an abundance of gypsum (naturally occurring and by-product chemical FGD and phospho gypsums) available worldwide, gypcrete-based building products, which are fully recyclable, offer significant environmental benefits.

### Metal

Metal is used as structural framework for larger buildings such as skyscrapers, or as an external surface covering. There are many types of metals used for building.

Metal figures quite prominently in prefabricated structures such as the Quonset hut, and can be seen used in most cosmopolitan cities. It requires a great deal of human labour to produce metal, especially in the large amounts needed for the building industries.

Corrosion is metal's prime enemy when it comes to longevity.

- Steel is a metal alloy whose major component is iron, and is the usual choice for metal structural building materials. It is strong, flexible, and if refined well and/or treated lasts a long time.
- The lower density and better corrosion resistance of aluminium alloys and tin sometimes overcome their greater cost.
- Copper is a valued building material because of its advantageous properties (see: Copper in architecture). These include corrosion resistance, durability, low thermal movement, light weight, radio frequency shielding, lightning protection, sustainability, recyclability, and a wide range of finishes.

  Copper is incorporated into roofing, flashing, gutters, downspouts, domes, spires, vaults, wall cladding, building expansion joints, and indoor design elements.
- Other metals used include chrome, gold, silver, and titanium. Titanium can be used for structural purposes, but it is much more expensive than steel. Chrome, gold, and silver are used as decoration, because these materials are expensive and lack structural qualities such as tensile strength or hardness.

### Plastics

The term "plastics" covers a range of synthetic or semi-synthetic organic condensation or polymerization products that can be molded or extruded into objects, films, or fibres. Their name is derived from the fact that in their semi-liquid state they are malleable, or have the property of plasticity.

Plastics vary immensely in heat tolerance, hardness, and resiliency. Combined with this adaptability, the general uniformity of composition and lightness of plastics ensures their use in almost all industrial applications today.

### Building Products

In the market place the term "building products" often refers to ready-made particles/sections, made from various materials, that are fitted in architectural hardware and decorative hardware parts of a building.

The list of building products excludes the building materials used to construct the building architecture and supporting fixtures, like windows, doors, cabinets, etc. Building products, rather, support and make building materials work in a modular fashion.

"Building products" may also refer to items used to put such hardware together, such as caulking, glues, paint, and anything else bought for the purpose of constructing a building.

# 2

# Deconstruction

In the context of physical construction, deconstruction is the selective dismantlement of building components, specifically for re-use, recycling, and waste management. It differs from demolition where a site is cleared of its building by the most expedient means. Deconstruction has also been defined as "construction in reverse". The process of dismantling structures is an ancient activity that has been revived by the growing field of sustainable, green method of building. Buildings, like everything, have a life-cycle. Deconstruction focuses on giving the materials within a building a new life once the building as a whole can no longer continue.

When buildings reach the end of their useful life, they are typically demolished and hauled to landfills. Building implosions or 'wrecking-ball' style demolitions are relatively inexpensive and offer a quick method of clearing sites for new structures. On the other hand, these methods create substantial amounts of waste. Components within old buildings may still be valuable, sometimes more valuable than at the time the building was constructed. Deconstruction is a method of harvesting what is commonly considered "waste" and reclaiming it into useful building material.

## Contribution to Sustainability

Deconstruction has strong ties to environmental sustainability. In addition to giving materials a new life cycle, deconstructing buildings helps to lower the need for virgin resources. This in turn leads to energy and emissions reductions from the refining and manufacture of new materials. As deconstruction is often done on a local level, many times on-site, energy and emissions are also saved in the transportation of materials. Deconstruction can potentially support communities by providing local jobs and renovated structures. Deconstruction work typically employs 3-6 workers for every one employed in a comparable demolition job. In addition, solid waste from conventional demolition is diverted from landfills. This is a major benefit because construction and demolition (C&D) waste accounts for approximately 20% of the solid waste stream.

### *Benefits of Avoiding Wood Waste*

In Canada, the CO2 Neutral Alliance has created a website with resources for regulators and municipalities, developers and contractors, business owners and operators, and individuals and households. Benefits for municipalities include:

- Reducing disposal costs where waste collection, hauling or disposal is supported by the tax base
- Establishing additional revenue streams
- Making existing landfills last longer
- Reducing greenhouse gas emissions caused by the decomposition of wood waste into methane from landfills
- Stimulating local economies with new industries and employment

Improving the local environment and overall sustainability of your community For every three square feet of deconstruction, enough lumber can be salvaged to build one square foot of new construction. At this rate, if deconstruction replaced residential demolition, the United States could generate enough recovered wood to construct 120,000 new affordable homes each year. The deconstruction of a typical 2,000-square-foot (190 $m^2$) wood frame home can yield 6,000 board feet of reusable lumber. Every year the United States buries about 33 million tons of wood-related construction and demolition debris in landfills. As anaerobic microorganisms decompose this wood, it will release about five million tons of carbon equivalent in the form of methane gas.

### *Typical Methods of Deconstruction*

Deconstruction is commonly separated into two categories; structural and non-structural. Non-structural deconstruction, also known as "soft-stripping", consists of reclaiming non-structural components, appliances, doors, windows, and finish materials. The reuse of these types of materials is commonplace and considered to be a mature market in many locales.

Structural deconstruction involves dismantling the structural components of a building. Traditionally this had only been performed to reclaim expensive or rare materials such as used brick, dimension stone, and extinct wood. In antiquity, it was common to raze stone buildings and reuse the stone; it was also common to steal stones from a building that was not being totally demolished: this is the literal meaning of the word *dilapidated.* Used brick and dimension limestone in particular have a long tradition of reuse due to their durability and colour changes over time. Recently, the rise of environmental awareness and sustainable building has made a much wider range of materials worthy of structural deconstruction. Low-end, commonplace materials such as dimensional lumber have become part of this newly emerging market.

The United States military has utilised structural deconstruction in many of its bases. The construction methods of barracks, among other base structures, are usually relatively simple. They typically contained large amounts of lumber and used minimal adhesives and finish-work. In addition, the buildings are often identical, making the process of deconstructing multiple buildings much easier. Many barracks were built during the era prior to WWII, and have aged to the point where they now need to be torn down. Deconstruction was deemed very practical due to the abundance of labour the military has access to and the value of the materials themselves.

Natural disasters, such as hurricanes, floods, tsunamis, and earthquakes often leave a vast amount of usable building materials in their wake. Structures that remain standing are often deconstructed to provide materials for rebuilding the region.

### Economic Potential

Deconstruction's economic viability varies from project to project. The amount of time and cost of labour are the main drawbacks. Harvesting materials from a structure can take weeks, whereas demolition may be completed in roughly a day. However, some of the costs, if not all, can be recovered. Reusing the materials in a new on-site structure, selling reclaimed materials, donating materials for income tax write-offs, and avoiding landfill "tipping fees" are all ways in which the cost of deconstruction can be made comparable to demolition.

Reclaiming the materials for a new on-site structure is the most economically and environmentally efficient option. Tipping fees and the costs of new materials are avoided; in addition, the transportation of the materials is non-existent. Selling the used materials or donating them to non-profit organisationsare another effective way of gaining capital. Donations to NPO's such as Habitat for Humanity's ReStore are tax deductible. Many times it is possible to claim the value to be half of what that particular material would cost new. When donating rare or antique components it is sometimes possible to claim a higher value than a comparable, brand-new material.

Value can also be added to new structures that are built by implementing reused materials. The United States Green Building Council's program entitled Leadership in Energy and Environmental Design (LEED) offers seven credits relating to reusing materials. (This accounts for seven out of a maximum sixty-nine credits) These include credits for building-shell reuse, material reuse, and diverting waste from landfills. Building shell-reuse is particularly appropriate for shells made of dimension stone.

Deconstruction is well suited to job training for the construction trades. Taking down a building is an excellent way for a worker to learn how to put a building up. This is vital for the economic recovery of inner-city communities. Unskilled and low-skilled workers can receive on-the-job training in use of basic carpentry tools and techniques, as well as learning teamwork, problem-solving, critical thinking and good work habits.

### *Process*

When choosing to deconstruct a building there are some important aspects that need to be taken into consideration. Developing a list of local contacts that are able to take used materials is an essential first step. These might include commercial architectural salvage businesses, reclamation yards, not-for-profit and social enterprise salvage warehouses, and dismantling contractors. Materials that cannot be salvaged may be recycled on-site or off-site, or taken to landfills. The next step involves identifying which, if any, are hazardous materials. Lead paint and asbestos are two substances in particular that need to be handled extremely cautiously and disposed of properly.

It is common practice, and common sense, to "soft-strip" the structure first; remove all appliances, windows, doors, and other finishing materials. These will account for a large percentage of the marketable components. After the non-structural deconstruction, structural is the next step. It is best to start at the roof and work down to the foundation.

Building components that are dismantled will need to be stored in a secure, dry location. This will protect them from water damage and theft. Once separated from the structure, materials can also be cleaned and/or refinished to increase value. Building an inventory list of the materials at hand will help determine where each item will be sent.

### Designing for Deconstruction (DfD)

"The end of the building's useful life generates a stream of used materials that can be reprocessed for new construction. The selection of materials for reuse or recycling should not start at the end of the building's life cycle, it should start at the design stage. Architects and engineers should keep the whole life cycle of the building in mind and select construction materials based on their capacity to be reused or recycled after the building has served its purpose."

An upstream approach to deconstruction can be implemented into buildings during their design process. This is a current trend in sustainable architecture. Often, simple construction methods combined with high-grade, durable materials work best for DfD structures. Separating layers of a building's infrastructure and making them visible can significantly simplify its deconstruction. Making components within systems separable also assists in being able to dismantle materials quickly and efficiently. This can be achieved by using mechanical fasteners such as bolts to connect parts. Allowing physical access to the fasteners is another needed aspect of this design. Also, it is important to use standardized materials and assemble them in a consistent manner throughout the project.

Some conventional construction methods and materials are difficult or impossible to deconstruct and should be avoided when designing for deconstruction. The use of nails and adhesives significantly slows down the deconstruction process and has a tendency to ruin otherwise reusable materials. Avoid hazardous materials altogether as they are detrimental to the natural environment and are non-reusable. Using mixed material grades makes the process of identifying pieces for resale difficult.

Deconstruction is important for more than just the end of a building's life-cycle. Buildings that have been designed with deconstruction in mind are often easier to maintain and adapt to new uses. Saving the shell of a building or adapting interior spaces to meet new needs ensures that new structures have a small environmental impact.

An alternative worth considering is modular building, like the Habitat 67 project in Montreal, Canada. This was a residential structure consisting of separate, functional apartments that could be put together in a variety of ways. As people moved in or out, the units could be reconfigured as desired.

## Major Appliance

A major appliance, or domestic appliance, is usually defined as a large machine which accomplishes some routine housekeeping task, which includes purposes such as cooking, or food preservation, whether in a household, institutional, commercial or industrial setting. An appliance is differentiated from a plumbing fixture because it uses an energy input for its operation other than water, generally using electricity or natural gas/propane. An object run by a watermill would also be considered an appliance. The term white goods or whiteware is also used for these items, primarily where British English is spoken, although definitions for the term "white goods" can differ. In the United States, the term *white goods* more commonly refers to linens rather than appliances.

Major appliances have become more technically complex from the control side recently with the introduction of the various Energy Labelling rules across the world. This has meant that the appliances have been forced to become more and more efficient leading to more accurate controllers in order to meet the regulations.

Major appliances are differentiated from small appliances because they are large, difficult to move, and generally fixed in place to some extent. They are often considered fixtures and part of real estate and as such they are often supplied to tenants as part of otherwise unfurnished rental properties. Another frequent characteristic of major appliances is that they may have substantial electricity requirements that necessitate special electrical wiring to supply higher current and/ or voltage than standard electrical outlets can deliver. This limits where they can be placed in a home.

### *Escalator*

An escalator is a moving staircase – a conveyor transport device for carrying people between floors of a building. The device consists of a motor-driven chain of individual, linked steps that move up or down on tracks, allowing the step treads to remain horizontal. Escalators are used around the world to move pedestrian traffic in places where elevators would be impractical. Principal areas of usage include department stores, shopping malls, airports, transit systems, convention centres, hotels, arenas, stadiums and public buildings.

The benefits of escalators are many. They have the capacity to move large numbers of people, and they can be placed in the same physical space as one might install a staircase. They have no waiting interval (except during very heavy traffic), they can be used to guide people toward main exits or special exhibits, and they may be weatherproofed for outdoor use. A non-functioning escalator can function as a normal staircase, whereas many other conveyances become useless when they break down.

### Design, Components, and Operation

***Operation and Layout:*** Escalators, like moving walkways, are often powered by constant-speed alternating currentmotors and move at approximately 1–2 feet (0.3–0.6 m) per second.

The typical angle of inclination of an escalator to the horizontal floor level is 30 degrees with a standard rise up to about 60 feet (18 m). Modern escalators have single-piece aluminium or stainless steel steps that move on a system of tracks in a continuous loop.

***Figure:*** *"Multiple parallel" layout*

Escalators have three typical configuration options: parallel (up and down escalators side by side or separated by a distance, seen often in metro stations and multilevel motion picture theaters), crisscross (minimizes space requirements by "stacking" escalators that go in one direction, frequently used in department stores or shopping centres), and multiple parallel (two or more escalators together that travel in one direction next to one or two escalators in the same bank that travel in the other direction).

***Figure:*** *"Crisscross" layout*

Escalators are required to have moving handrails that keep pace with the movement of the steps. This helps riders steady themselves, especially when stepping onto the moving stairs. Occasionally, a handrail will move at a slightly different speed from the steps, causing it to "creep" slowly forward or backward relative to the steps. Riders may find this effect disconcerting and this problem should be fixed.

The direction of escalator movement (up or down) can be permanently set, or be controlled by personnel according to the predominant flow of the crowd, or be controlled automatically. In some setups, direction is controlled automatically by whoever arrives first, whether at the bottom or at the top (the system is programmed so that the direction is not reversed while a passenger is on the escalator).

### Design and Layout Considerations

A number of factors affect escalator design, including physical requirements, location, traffic patterns, safety considerations, and aesthetic preferences. Foremost, physical factors like the vertical and horizontal distance to be spanned must be considered.

These factors will determine the length and pitch of the escalator. The building infrastructure must be able to support the heavy components. The escalator should be located where it can be easily seen by the general public. In department stores, customers should be able to view the merchandise easily. Furthermore, up and down escalator traffic should be physically separated and should not lead into confined spaces.

Traffic patterns must also be anticipated. In some buildings, the objective is simply to move people from one floor to another, but in others there may be a more specific requirement, such as funneling visitors towards a main exit or exhibit. The escalators must be designed to carry the required number of passengers.

For example, a single-width escalator travelling at about 1.5 feet (0.5 m) per second can move about 2000 people per hour. The carrying capacity of an escalator system must match the expected peak traffic demand, presuming that passengers ride single file. This is crucial if there are sudden increases in the number of riders. For example, escalators at stations must be designed to cater for the peak traffic flow discharged from a train, without causing excessive bunching at the escalator entrance.

In this regard, escalators help in controlling the flow of people. For example, an escalator to an exit effectively discourages most people from using it as an entrance, and may reduce security concerns. Similarly, escalators often are used as the exit from airport security checkpoints. Such an egress point would generally be staffed to prevent its use as an entrance, as well.

It is preferred that staircases be located adjacent to the escalator if the escalator is the primary means of transport between floors. It may also be necessary to provide an elevator lift near the escalator for wheelchairs and disabled people. Finally, consideration should be given to the aesthetics of the escalator.

## Model Sizes and Other Specifications

| *Escalator step widths and energy usage* | | | | |
|---|---|---|---|---|
| ***Size*** | ***Width (between balustrade panels)*** | ***Single-step capacity*** | ***Applications*** | ***Power consumption*** |
| Very small | 400 mm (16 in) | One passenger, with feet together1989 | A rare historic design found mostly in older department stores | 3.7 kW (5 hp) |
| Small | 600 mm (24 in) | One passenger | Low-volume sites, uppermost levels of department stores, when space is limited | 3.7 kW (5 hp) |
| Medium | 800 mm (31 in) | One passenger + one package or one piece of luggage | Shopping malls, department stores, smaller airports | 7.5 kW (10 hp) |
| Large | 1,000 mm (39 in) | Two passengers – one may walk past another | Mainstay of metro systems, larger airports, train stations, some retail usage | 7.5 kW (10 hp) |

## Components

Escalator truss connects to the landing platform (lower left).

Also visible: exposed drive gears (centre) for steps and handrail drive (left).

***Figure:*** *View of escalator steps on continuous chain*

***Figure:*** *An escalator equipped with a "bellows" handrail. The bracelets are coloured grey, with occasional strings of black ones to appear as moving spacers.*

### Landing Platforms

These two platforms house the curved sections of the tracks, as well as the gears and motors that drive the stairs. The top platform contains the motor assembly and the main drive gear, while the bottom holds the step return idler sprockets. These sections also anchor the ends of the escalator truss. In addition, the platforms contain a floor plate and a combplate.

The floor plate provides a place for the passengers to stand before they step onto the moving stairs. This plate is flush with the finished floor and is either hinged or removable to allow easy access to the machinery below. The combplate is the piece between the stationary floor plate and the moving step. It is so named because its edge has a series of cleats that resemble the teeth of a comb. These teeth mesh with matching cleats on the edges of the steps. This design is necessary to minimize the gap between the stair and the landing, which helps prevent objects from getting caught in the gap.

### Truss

The truss is a hollow metal structure that bridges the lower and upper landings. It is composed of two side sections joined together with cross braces across the bottom and just below the top. The ends of the truss are attached to the top and bottom landing platforms via steel or concrete supports. The truss carries all the straight track sections connecting the upper and lower sections.

### Tracks

The track system is built into the truss to guide the step chain, which continuously pulls the steps from the bottom platform and back to the top in an endless loop. There are actually two tracks: one for the front wheels of the steps (called the step-wheel track) and one for the back wheels of the steps (called the trailer-wheel track). The relative positions of these tracks cause the steps to form a staircase as they move out from under the combplate. Along the straight section of the truss the tracks are at their maximum distance apart. This configuration forces the back of one step to be at a 90-degree angle relative to the step behind it. This right angle bends the steps into a shape resembling a staircase. At the top and bottom of the escalator, the two tracks converge so that the front and back wheels of the steps are almost in a straight line. This causes the stairs to lay in a flat sheetlike arrangement, one after another, so they can easily travel around the bend in the curved section of track. The tracks carry the steps down along the underside of the truss until they reach the bottom landing, where they pass through another curved section of track before exiting the bottom landing. At this point the tracks separate and the steps once again assume a staircase configuration. This cycle is repeated continually as the steps are pulled from bottom to top and back to the bottom again.

### Steps

The steps themselves are solid, one piece, die-cast aluminium or steel. Yellow demarcation lines may be added to clearly indicate their edges. In most escalator models manufactured after 1950, both the riser and the tread of each step is cleated (given a ribbed appearance) with comblike protrusions that mesh with the combplates on the top and bottom platforms and the succeeding steps in the chain. Seeberger-or "step-type" escalators featured flat treads and smooth risers; other escalator models have cleated treads and smooth risers. The steps are linked by a continuous metal chain that forms a closed loop. The front and back edges of the steps are each connected to two wheels. The rear wheels are set further apart to fit into the back track and the front wheels have shorter axles to fit into the narrower front track. As described above, the position of the tracks controls the orientation of the steps.

### *Handrail*

The handrail provides a convenient handhold for passengers while they are riding the escalator. In an escalator, the handrail is pulled along its track by a chain that is connected to the main drive gear by a series of pulleys. It is constructed of four distinct sections. At the centre of the handrail is a "slider", also known as a "glider ply", which is a layer of a cotton or synthetic textile. The purpose of the slider layer is to allow the handrail to move smoothly along its track. The next layer, known as the "tension member", consists of either steel cable or flat steel tape, and provides the handrail with tensile strength and flexibility. On top of tension member are the inner construction components, which are made of chemically treated rubber designed to prevent the layers from separating. Finally, the outer layer—the only part that passengers actually see—is the cover, which is a blend of synthetic polymers and rubber. This cover is designed to resist degradation from environmental conditions, mechanical wear and tear, and human vandalism.

In the factory, handrails are constructed by feeding rubber through a computer-controlled extrusion machine to produce layers of the required size and type in order to match specific orders. The component layers of fabric, rubber, and steel are shaped by skilled workers before being fed into the presses, where they are fused together.

In the mid-twentieth century, some handrail designs consisted of a rubber bellows, with rings of smooth metal cladding called "bracelets" placed between each coil. This gave the handrail a rigid yet flexible feel. Additionally, each bellows section was no more than a few feet long, so if part of the handrail was damaged, only the bad segment needed to be replaced. These forms of handrail have largely been replaced with conventional fabric-and-rubber railings.

### *Safety*

Safety is also major concern in escalator design. In India where women wear saris, there are heavy chances of getting the *pallu* entangled in the escalator. As a result, a special *sari guard* is built into most escalators. There is a risk of feet injuries for children wearing footwear such as Crocs and flip-flops that might get caught in escalator mechanisms. This was due to the softness of the shoe's material combined with the smaller size of children's feet.

Fire protection of an escalator floor opening may be provided by adding automatic sprinklers or fireproof shutters to the opening, or by installing the escalator in an enclosed fire-protected hall. To limit the danger of overheating, ventilation for the spaces that contain the motors and gears must be provided.

### Accidents and Litigation

***Accidents:*** There have been reports of people falling off a moving escalator or getting their shoe stuck in part of the escalator; shoe laces are a hazard when loose. Some accidents are caused by improper or unsafe use such as riding the hand rails or by escalator spinning. A few fatal accidents are:

- Eight people died and 30 more were injured on February 17, 1982, when an escalator collapsed on the Moscow Metro. Wrongly set up service brakes were later blamed for the accident.
- 31 people died after a fire, begun in the undercarriage of an MH-type Otis escalator, exploded into the ticketing hall at King's Cross St. Pancras station in 1987.
- On December 13, 1999, 8-year-old Jyotsna Jethani was killed at New Delhi's international airport. Jethani fell into a gaping hole that resulted from improper maintenance.
- On June 15, 2002, Andrea Albright, a 24-year-old J.C. Penney employee in Columbia, Maryland, was injured while riding the store's escalator, and died from her injuries 10 days later.
- On July 2, 2003, 30 people were injured when an escalator at Coors Field in Denver malfunctioned, and one victim eventually had to have a leg amputated. The accident is believed to have been caused by a combination of overcrowding and an improperly installed safety device.
- On December 31, 2004, escalators at the Taipei City Hall Station kept moving commuters onto the overcrowded island platform. A woman whose hair got caught in the escalator received 20 stitches to the scalp.
- Francisco Portillo died after being strangled when his sweatshirt got caught in an escalator at the Porter SquareMBTA station in Cambridge, Massachusetts on February 21, 2005. He was allegedly drunk at the time.
- On September 13, 2008, an 11-year old boy died after falling off an escalator in Lyngdal, Norway. On April 20, 2009, a teenage boy died after getting very serious skull injuries after falling off an escalator in Falun, Sweden. On June 26, 2009, a man died after falling off an escalator in Helsingborg, Sweden. All three were riding the handrail.

- On July 6, 2011, a 13-year old boy was killed and several people were injured when they were thrown off an escalator that suddenly changed direction in a Beijing subway station. Officials identified the manufacturers of the escalator as a possible reason for the accident. The escalator was built by Otis and was still in its guarantee period.
- On August 3, 2011, 35 year old deputy stationmaster Kuo Cheng-chieh was killed after being ingested into an escalator. He was carrying a long board used for flood defences on an upward-moving escalator at Dingxi Station in New Taipei City, Taiwan. The board became wedged between the ceiling and the step Kuo was standing on, the step collapsed under the load, and the victim fell into the machinery. The event didn't activate any safety mechanisms and the escalator continued to run.
- On August 30, 2012, a man was killed when he fell 5 stories off an escalator rail at Reliant Stadium during a Texans game.
- On January 7, 2013, at least 6 people were injured when the crowded escalator at the Exchange Place PATH station suddenly reversed direction.

### Lessons of the King's Cross Fire

The King's Cross fire of 1987 illustrated the demanding nature of escalator upkeep and the devices' propensity to collect "fluff" when not properly maintained. In the official inquiry that followed, the Fennell Report, it was determined that the fire started slowly, smoldered virtually undetected for a time, then exploded into the ticketing hall above in a phenomenon known as the "trench effect." In the escalators' undercarriage, approximately 8,800 kilograms (19,400 lb) of accumulated detritus acted as a wick to a neglected buildup of interior lubricants; wood veneers, paper and plastic advertisements, solvent-based paint, plywood in the ticket hall, and melamine combustion added to the impact of the calamity.

As a result of the report, older wooden escalators were removed from service in the London Underground. Additionally, sections of the London Underground that were actually below ground were made nonsmoking; eventually the whole system became a smoke-free zone.

### Litigation

In the 1930s, at least one suit was filed against a department store, alleging that its escalators posed an attractive nuisance, responsible for a child's injury. These cases were almost always dismissed. Moreover, continual updating of escalator safety codes facilitated increased levels of consumer safety as well as a reduction in court cases.

### Legislation and Escalators

***United States:*** Despite their considerable scope, two Congressional Acts, the Rehabilitation Act of 1973 and the Americans with Disabilities Act of 1990 (ADA), did not directly affect escalators or their public installations. Since Section 504 of the Rehabilitation Act included public transportation systems, for a few years, the United States Department of Transportation considered designs to retrofit existing escalators for wheelchair access. Nonetheless, Foster-Miller Associates' 1980 plan, *Escalator Modification for the Handicapped* was ultimately ignored in favour of increased elevator installations in subway systems. Likewise, the ADA provided more accessibility options, but expressly excluded escalators as "accessible means of egress," advocating neither their removal nor retention in public structures.

### Codes and Regulation

In the United States and Canada, new escalators must abide by ASMEA17.1 standards, and old/historic escalators must conform to the safety guidelines of ASMEA17.3. In Europe, the escalator safety code is EN115.

### Key Safety Features

To enhance passenger safety, newer models of escalators are equipped with one or more of the following safety implementations, as per ASMEA17.1 code:

- Antislide devices: Raised circular objects that often stud the escalator balustrade. Sometimes informally called "hockey pucks" due to their appearance, their purpose is to prevent objects (and people) from precipitously sliding down the otherwise smooth metallic surface.
- Combplate impact switches: Stop the escalator if a foreign object gets caught between the steps and the combplate on either end.
- Deflector brush: A long continuous brush made of stiff bristles running up the sides of the escalator just above the step level. This helps deflect garments, shoes, and other items away from the gap between the moving steps and the skirt board.

- Emergency stop button: At each end of the escalator (in some models, also on the balustrade), a large red button can be pressed to stop the device in the event of an emergency. Typically, an alarmed transparent plastic guardplate covers the button; restarting requires turning a key.
- Extended balustrades: Allows riders to grasp the handrail before setting foot on an escalator, to ease customer comfort and stability/equilibrium. (The effect is similar to the flat steps described below.)
- Flat steps: Like a moving walkway, the first two or three steps at either end of the escalator are flat. This gives the passenger extra time to orient him/herself when boarding, and more time to maintain balance when exiting. Longer escalators often have four or more flat steps.
- Handrail inlet switches: Sensors located at the bottom and top of the unit that guard the handrail termini. If something gets caught in these locations, a hard fault is generated in the controller, and the escalator shuts down automatically.
- Handrail speed sensors: These sensors are usually optical, and monitor how fast the handrail moves. If the sensor notices a speed difference between the handrail and the steps, it sounds an alarm, pauses, and then automatically stops the escalator. In these situations, the escalator must be serviced by authorized personnel before returning to an operable state.
- Missing step detectors: Depending on the manufacturer and model, this sensor is either optical or physical. When a missing step is detected, the escalator automatically shuts down.
- Raised step edges: In some models, a difference in tread height is utilised to keep passengers' feet from the skirt board.
- Safety instructions: A sign, typically posted on both escalator newels at the entrance landing platform. In some situations, safety precautions are posted on walls near the escalator, included on freestanding signs, or—as in some models—printed on the riser surface itself.
- Sensor switch: In automatic-start/stop escalators, this sensor automatically engages the escalator motion when a rider is detected on the first step of the entrance landing platform, and stops the escalator when there are no riders on the unit.
- Step demarcation lights: Either fluorescent or LED lights (traditionally green in colour) located inside the truss. The

illumination between the steps improves the passengers' awareness of the step divisions.

- Step demarcation lines: In order to clearly delineate the edges of each individual step, manufacturers offer steps trimmed in yellow, either painted or with plastic inserts.

### History

Inventors and manufacturers:

***Nathan Ames:*** Nathan Ames, a patent solicitor from Saugus, Massachusetts, is credited with patenting the first "escalator" in 1859, despite the fact that no working model of his design was ever built. His invention, the "revolving stairs", is largely speculative and the patent specifications indicate that he had no preference for materials or potential use (he noted that steps could be upholstered or made of wood, and suggested that the units might benefit the infirm within a household use), though the mechanization was suggested to run either by manual or hydraulic power.

### Leamon Souder

In 1889, Leamon Souder successfully patented the "stairway", an escalator-type device that featured a "series of steps and links jointed to each other". No model was ever built. This was the first of at least four escalator-style patents issued to Souder, including two for spiral designs (U. S. Patent Nos. 723,325 and 792,623).

### Jesse Wilford Reno, George A. Wheeler, and Charles Seeberger

In 1892, Jesse W. Reno patented the "Endless Conveyor or Elevator." A few months after Reno's patent was approved, George A. Wheeler patented his ideas for a more recognisable moving staircase, though it was never built. Wheeler's patents were bought by Charles Seeberger; some features of Wheeler's designs were incorporated in Seeberger's prototype built by the Otis Elevator Company in 1899.

Reno, a graduate of Lehigh University, produced the first working escalator (he actually called it the "inclined elevator") and installed it alongside the Old Iron Pier at Coney Island, New York City in 1896. This particular device was little more than an inclined belt with cast-iron slats or cleats on the surface for traction, and travelled along a 25° incline. A few months later, the same prototype was used for a month long trial period on the Manhattan side of the Brooklyn Bridge. Reno eventually joined forces with Otis, and retired once his patents were purchased outright. Some Reno-type escalators were still being used in the Boston subway until construction for the Big Dig precipitated their removal. The Smithsonian Institution considered re-assembling one of these historic units from 1914 in their collection of Americana, but "logistics and reassembly costs won out over nostalgia", and the project was discarded.

Around May 1895, Charles Seeberger began drawings on a form of escalator similar to those patented by Wheeler in 1892. This device actually consisted of flat, moving stairs, not unlike the escalators of today, except for one important detail: the step surface was smooth, with no comb effect to safely guide the rider's feet off at the ends. Instead, the passenger had to step off sideways. To facilitate this, at the top or bottom of the escalator the steps continued moving horizontally beyond the end of the handrail (like a miniature moving sidewalk) until they disappeared under a triangular "divider" which guided the passenger to either side. Seeberger teamed with Otis in 1899, and together they produced the first commercial escalator which won the first prize at the Paris 1900 *Exposition Universelle* in France. Also on display at the *Exposition* were Reno's inclined elevator, a similar model by James M. Dodge and the Link Belt Machinery Co., and two different devices by French manufacturers Hallé and Piat.

### Early European Manufacturers: Hallé, Hocquardt, and Piat

Piat installed its "stepless" escalator in Harrods Knightsbridge store on Wednesday, November 16, 1898, though the company relinquished its patent rights to the department store. Noted by Bill Lancaster in *The Department Store: a Social History*, "customers unnerved by the experience were revived by shopmen dispensing free smelling salts and cognac." The Harrods unit was a continuous leather belt made of "224 pieces . . . strongly linked together travelling in an upward direction," and was the first "moving staircase" in England.

Hocquardt received European patent rights for the *Fahrtreppe* in 1906. After the *Exposition*, Hallé continued to sell its escalator device in Europe, but was eventually eclipsed in sales by other major manufacturers.

### Major Competitors and Product Nomenclature

In the first half of the twentieth century, several manufacturers developed their own escalator products, though they had to market their devices under different names, due to Otis' hold on the trademark rights to the word "escalator." New York-based Peelle Company called their models the *Motorstair*, and Westinghouse called their model an

*Electric Stairway*. The Toledo-based Haughton Elevator company referred to their product as simply *Moving Stairs*.

### Manufacturing Mergers and Buyouts: the Playing Field Narrows

Kone and Schindler introduced their first escalator models several decades after the Otis Elevator Co., but grew to dominance in the field over time. Today, they, Mitsubishi, and ThyssenKrupp are Otis' primary rivals. Schindler now stands as the largest maker of escalators and second largest maker of elevators in the world, though their first escalator installation did not occur until 1936. In 1979, the company entered the United States market by purchasing Haughton Elevator; nine years later, Schindler assumed control of the North American escalator/elevator operations of Westinghouse.

Kone expanded internationally by acquisition in the 1970s, buying out Swedish elevator manufacturer Asea-Graham, and purchasing other minor French, German, and Austrian elevator makers before assuming control of Westinghouse's European elevator business. As the last "big four" manufacturers held on to the escalator market, KONE first acquired Montgomery Elevator Company, then took control of Germany's Orenstein & Koppel *Rolltreppen*.

### Model Development and Design Types

***"Cleat-Type" Escalators:*** Jesse Reno's escalators did not resemble modern escalators too closely. Passengers' feet tilted upward at an angle, and the treads consisted of cleated metal (initially) or wood (later models). Reno worked on his own for several years, gaining success with installations from Toronto to Cape Town, South Africa. Similar units of the day by other manufacturers resembled conveyor belts more than moving staircases. For a time, Otis Elevator sold Reno's escalators as their own "cleat-type" escalators.

### "Step-Type" Escalators

Seeberger's model, bought by Otis, clearly became the first "step-type" escalator, so called after its visual likeness to steps on a regular staircase. The company later combined the best aspects of both inventions (guiding slats and flat steps) and in 1921 produced an escalator similar to the type used today: they called it the "L-type" escalator. It was succeeded by the "M-type", the "O-type", and current models by Otis such as the "NCE-type" escalator.

### Spiral Escalators

Jese Reno, in addition to being known for the first "practical" escalator in public use, also bears the unique distinction of designing the very first escalators installed in any underground subway system – a single spiral escalator at Holloway Road tube station in London in 1906. The experimental device never saw public use, and was forgotten for several decades. The remains of this are now in the London Transport Museum's depot in Acton. Also the first fully operational spiral escalator, Reno's design was nonetheless only one in a series of several similar proposed contraptions. Souder patented two spiral designs, Wheeler drafted spiral stairway plans in 1905, Seeberger devised at least two different spiral units between 1906 and 1911 (including an unrealised arrangement for the London Underground), and Gilbert Luna obtained West German, Japanese, and United States patents for his version of a spiral escalator by 1973. When interviewed for the *Los Angeles Times* that year, Luna was in the process of soliciting "major firms" for acquisition of his patents and company, but statistics are unclear on the outcome of his endeavours in that regard.

The Mitsubishi Electric Corporation was most successful in its development of "spiral" (more "curve" than true spiral) escalators, and has sold them exclusively since the mid-1980s. The world's first "practical" spiral escalator—a Mitsubishi model—was installed in Osaka, Japan, in 1985. In use, a major planning advantage presented by spiral escalators is that they take up much less horizontal floor space than traditional units, which frequently house large machine rooms underneath the truss.

Helixator, an experimental spiral escalator design which currently exists as a prototype scale model, could further reduce the floor space. The design has several innovations that offer the possibility of a continuous helix. Driven by a linear motor instead of a chain system, the Helixator spreads the force evenly along the escalator path, avoiding excessive force on the top chain links and removing geometry, length and height limits.

### Etymology

Several authors and historians have contributed their own differing interpretations of the source of the word "escalator", and some degree of misinformation has heretofore proliferated on the Internet. For reference, contradictory citations by seven separate individuals, including the Otis Elevator Company itself, are provided below.

### Name Development and Original Intentions

Charles Seeberger created the word "escalator" in 1900, to coincide with his device's debut at the *Exposition Universelle*. According to his own account, in 1895, his legal counsel advised him to name his new invention, and he then set out to devise a title for it on his own. As evidenced in Seeberger's own handwritten documents, archived at the Otis Elevator Company headquarters in Farmington, Connecticut, the inventor consulted "a Latin lexicon" and "adopted as the root of the new word, 'Scala'; as a prefix, 'E' and as a suffix, 'Tor.'" His own rough translation of the word thus created was "means of traversing from", and he intended for the word to be pronounced, "es' [1]kæl' Y' tYr" (es-CAL-a-tor).

"Escalator" was not a combination of other French or Greek words, and was never a derivative of "elevator" in the original sense, which means "one who raises up, a deliverer" in Latin. Similarly, the root word "*scala*" does not mean "a flight of steps", but is defined by Lewis and Short's *A Latin Dictionary* as the singular form of the plural noun "*scalae*", which denotes any of the following: "a flight of steps or stairs, a staircase; a ladder, [or] a scaling-ladder."

The alleged intended capitalization of "escalator" is likewise a topic of debate. Seeberger's trademark application lists the word not only with the "E" but also with all of the letters capitalized (in two different instances), and he specifies that, "any other form and character of type may be employed . . . without altering in any essential manner the character of [the] trade-mark." That his initial specifications are ostensibly inconsistent, and since Otis Elevator Co. advertisements so frequently capitalized all of the letters in the word, suppositions about the "capital 'e'" are difficult to formulate.

### *Derivatives of 'Escalator'*

The verb "escalate" originated in 1922, and has two uses, the primary: "to climb or reach by means of an escalator" or "to travel on an escalator", and the secondary: "to increase or develop by successive stages; *spec.* to develop from 'conventional' warfare into nuclear warfare." The latter definition was first printed in the *Manchester Guardian* in 1959, but grew to prominent use during the late 1960s and early 1970s.

### *Loss of Trademark Rights*

In 1950, the landmark case *Haughton Elevator Co. v. Seeberger* precipitated the end of Otis's reign over exclusive use of the word "escalator", and simultaneously created a cautionary study for companies and individuals interested in trademark retention. Confirming the contention of the Examiner of Trademark Interferences, Assistant Commissioner of Patents Murphy's decision rejected the

Otis Elevator Company's appeal to keep their trademark intact, and noted that "the term 'escalator' is recognised by the general public as the name for a moving stairway and not the source thereof", observing that the Otis Elevator Co. had "used the term as a generic descriptive term...in a number of patents which [had] been issued to them and...in their advertising matter." All trademark protections were removed from the word "escalator", the term was officially genericized, and it fell into the public domain.

### Primary Uses and Application

***Department Stores/Shopping:*** As noted above, a few escalator types were installed in major department stores (including Harrods) before the *Expo*. Escalators proved instrumental in the layout and design of shopping venues in the twentieth century. By 1898, the first of Reno's "inclined elevators" were incorporated into the Bloomingdale Bros. store at Third Avenue and 59th Street. This was the first retail application of the devices in the US, and no small coincidence, considering that Reno's primary financier was Lyman Bloomingdale, co-owner of the department store with brother Joseph Bloomingdale.

### Public Transportation

The first "standard" escalator installed on the London Underground was a Seeberger model at Earls Court. Noted above, London's Underground installed a rare spiral escalator designed by Reno, William Henry Aston and Scott Kietzman for the Holloway Road Underground station in 1906; it was run for a short time but was taken out of service the same day it debuted. The older lines of the London Underground had many escalators with wooden treads (ca. 1930s) until they were rapidly replaced following the King's Cross fire, noted above.

### Other Applications

***Factories and Other Industrial Production Environments:*** In 1905, the American Woolen Company's Wood Mill in Lawrence, Massachusetts (then "the largest single worsted mill in the world") utilised Otis' Seeberger-type "reversible" escalators to carry its workers between floors four times a day. The machines did not run all day: rather, escalators ran solely to transport employees to/from midday meals and in/out of the mill. In its advertising, Otis Elevator Company hailed this unconventional use for its unique benefits to both workers and owners: "The profitable and practicable feature of the Escalator, from the viewpoint of the owner, is the increased efficiency of each operator due to the elimination of stair climbing."

### Military Use

In San Francisco, an escalator at Hunters Point Naval Shipyard was used to convey personnel between the first and third floors. At the time of its construction in 1948, it was touted thus: "[it has the] highest lift of any industrial building in the world. It rises 42 feet (12.8 metres)."

Escalators were also utilised on aircraft carriers such as the USS *Hornet* (CV-12), to transport pilots from "ready rooms" to the flight deck.

### Extant Historic Escalator Models

A few notable examples of historic escalators still in operation are:

### Australia

- Town Hall Railway Station, Sydney, Australia
- Wynyard Railway Station, Sydney, Australia

### Europe

- St. Anna Pedestrian Tunnel underneath the Schelde in Antwerp, Belgium. This tunnel was opened in 1933.
- Tyne Cyclist and Pedestrian Tunnel, Tyne and Wear, England

These escalators, manufactured by Waygood Otis in 1951, were "believed to be the longest single lift escalators in the world", at the time of installation. Presumably the first escalators in Britain designed specifically for cyclists, they were also the longest in the United Kingdom. At most, they may be the longest extant wooden escalators in the world, though they are no longer in working order.

- Baumanskayametro station in Moscow, Russia contains the oldest (1944) escalator of tonnel type still in use, with length of 65 m (213 ft) and vertical rise of 32.5 m (107 ft).
- Greenford Station in Greenford, England contains the oldest (1947) escalator still in use on the London Underground service. This was the first above ground station to install an escalator to its high levels and one of the few (if not the only) wooden escalators not to be replaced after the fire at Kings Cross station.

### North America

- Macy's Herald Square department store, Otis L-type units with wood treads and replacement metal treads, New York,

New York (only in escalators heading upwards from the first to the seventh stories)

- Kaufmann's department store (now Macy's), two 16-inch (406 mm) Otis L-type units with original floorplates, several 40-inch (1,016 mm) Otis escalators ca. 1950s, Pittsburgh, Pennsylvania
- Westfield San Francisco Centre (formerly The Emporium), chrome-and-glass escalator by Eleanor LeMaire for Otis, San Francisco, California
- The Land Pavilion-Epcot, Glass escalator by Otis. Has stairs next to them. Not single file but is separated
- Tokyo Sesame Place-Glass escalator with Big Bird's mouth as the entrance. First escalator in a theme park (now closed)

### Escalators: Superlatives

Longest systems:

- Central-Mid-Levels escalator: in Hong Kong, tens of thousands of commuters travel each work day between Central, the central business district, and the Mid-levels, a residential district hundreds of feet uphill, using this long distance system of escalators and moving walkways. It is the world's longest outdoor escalator *system* (not a single escalator span), at a total length of 2,600 feet (790 m). It goes only one way at a time; the direction reverses depending on rush hour traffic direction.
- In December 2011, a network of six escalators of 1,260 feet (380 m) length, equivalent to 28 stories high, was opened in Medellin, Colombia, offering the 12,000 residents of Comuna 13 a six-minute ride to the city centre compared to their previous 35-minute climb on foot.
- Cascade, Yerevan: an escalator system of 774 ft (236 m) length and 387 ft (118 m) height.
- Ocean Park, Hong Kong: a long escalator system connecting two parts of the Park, with an overall length of 730 feet (220 m).

### Longest Individual Escalators

World:

- The longest escalators in the world are installed in deep underground stations of the Saint Petersburg Metro. Three stations have escalators up to 132 m (433 ft) long and 66 m (217 ft) high: Ploshchad Lenina, Chernyshevskaya, and Admiralteyskaya.

- The longest *freestanding* (supported only at the ends) escalator in the world is inside CNN Centre's atrium in Atlanta, Georgia. It rises 8 stories and is 205 feet (62 m) long. Originally built as the entrance to the amusement park *The World of Sid and Marty Krofft*, the escalator is now used for CNN studio tours.

Asia:

- The longest escalator in Bangkok, Thailand and Southeast Asia is in Bangkok Metro's Si Lom Station. It connects the concourse level with platform 1 which in turn connects to Hua Lam Phong. It is 43 m (141 ft) in length and 21.5 m (71 ft) in depth.

Australia:

- The longest set of single-span uninterrupted escalators in the southern hemisphere is at Parliament underground railway station in Melbourne.

Europe:

Several metro or subway systems in Central and Eastern Europe feature very long escalators.

- The Kiev MetroKreschatik station's lower-level second exit escalator (a type ЁÒ-2, circa 1965), lifts riders 65.7 metres (216 ft), or 743 steps, up a 131.4-metre (431 ft) – long incline.
- In the Park Pobedy station of the Moscow Metro, the escalators are 126.8 m (416 ft) or 740 steps long, and 63.4 m (208 ft) high. It takes three minutes to transit.
- The longest escalator in Prague, and in European Union, is at the NámìstíMíru station at 285 feet (87 m) long and 142.5 feet (43.4 m) high.
- The longest escalator of a European shopping mall is at MyZeil, Frankfurt, Germany, with a length of 150 feet (46 m).
- The tallest escalator on the London Underground system is at Angel station with a length of 200 feet (61 m), and a vertical rise of 90 feet (27 m).
- The longest wooden escalators in the United Kingdom are at the Tyne Tunnel, with a length of 197 feet (60 m).
- The longest escalators in Western Europe are in the Stockholm Metro, at Västraskogen with a length of 220 feet (67 m) and

in Helsinki Metro, at Kamppi station with a length of 210 feet (64 m).

- The largest "single truss escalator" is in the Bentall Centre in Kingston upon Thames in Greater London, UK. It connects the ground floor with the second floor with only top and bottom supports.

### *North and South America*

- The longest set of single-span uninterrupted escalators in the Western Hemisphere is at the Wheaton station of the Washington Metro system. They are 230 feet (70 m) long with a vertical rise of 115 feet (35 m), and take what is variously described as 2 minutes and 45 seconds or nearly three-and-a-half minutes, to ascend or descend without walking.

### *Shortest Examples*

According to Guinness, the shortest escalator in the world is in the Okadaya Mores shopping mall in Kawasaki, Japan. Its vertical rise is only 32.8 inches (83 cm).

## Electrical Wiring

Electrical wiring in general refers to insulated conductors used to carry electricity, and associated devices. This article describes general aspects of electrical wiring as used to provide power in buildings and structures, commonly referred to as building wiring. This article is intended to describe common features of electrical wiring that may apply worldwide. For information regarding specific national electrical codes, refer to the articles mentioned in the next section. Separate articles cover long-distance electric power transmission and electric power distribution.

### *Wiring Safety Codes*

Wiring safety codes are intended to protect people and property from electrical shock and fire hazards. Regulations may be established by city, county, provincial/state or national legislation, usually by adopting a model code (with or without local amendments) produced by a technical standards-setting organisation, or by a national standard electrical code.

Electrical codes arose in the 1880s with the commercial introduction of electrical power. Many conflicting standards existed for the selection of wire sizes and other design rules for electrical installations.

The first electrical codes in the United States originated in New York in 1881 to regulate installations of electric lighting. Since 1897 the US National Fire Protection Association, a private non-profit association formed by insurance companies, has published the *National Electrical Code* (NEC). States, counties or cities often include the NEC in their local building codes by reference along with local differences. The NEC is modified every three years. It is a consensus code considering suggestions from interested parties. The proposals are studied by committees of engineers, tradesmen, manufacturer representatives, fire fighters, and other invitees.

Since 1927, the Canadian Standards Association (CSA) has produced the Canadian *Safety Standard for Electrical Installations*, which is the basis for provincial electrical codes. The CSA also produces the Canadian Electrical Code, the 2006 edition of which references IEC 60364 (*Electrical Installations for Buildings*) and states that the code addresses the fundamental principles of electrical protection in Section 131. The Canadian code reprints Chapter 13 of IEC 60364, but there are no numerical criteria listed in that chapter to assess the adequacy of any electrical installation.

Although the US and Canadian national standards deal with the same physical phenomena and broadly similar objectives, they differ occasionally in technical detail. As part of the North American Free Trade Agreement (NAFTA) program, US and Canadian standards are slowly converging toward each other, in a process known as harmonization. In European countries, an attempt has been made to harmonize national wiring standards in an IEC standard, IEC 60364 *Electrical Installations for Buildings*. Hence national standards follow an identical system of sections and chapters. However, this standard is not written in such language that it can readily be adopted as a national wiring code. Neither is it designed for field use by electrical tradesmen and inspectors for testing compliance with national wiring standards. By contrast, national codes, such as the NEC or CSA C22.1, generally exemplify the common objectives of IEC 60364, but provide specific rules in a form that allows for guidance of those installing and inspecting electrical systems.

In Germany, DKE (the German Commission for Electrical, Electronic and Information Technologies of DIN and VDE) is the organisation responsible for the promulgation of electrical standards and safety specifications. DIN VDE 0100 is the German wiring regulations document harmonised with IEC 60364. In the United Kingdom, wiring installations are regulated by the Institution of

Engineering and Technology *Requirements for Electrical Installations: IEE Wiring Regulations, BS 7671: 2008,* which are harmonised with IEC 60364. The 17th edition (issued in January 2008) includes new sections for microgeneration and solar photovoltaic systems. The first edition was published in 1882.

In Australia and New Zealand, the AS/NZS 3000 standard, commonly known as the "wiring rules", specifies requirements for the selection and installation of electrical equipment, and the design and testing of such installations. The standard is mandatory in both New Zealand and Australia; therefore, all electrical work covered by the standard must comply. The international standard wire sizes are given in the IEC 60228 standard of the International Electrotechnical Commission. In North America, the American Wire Gauge standard for wire sizes is used.

## Colour Code

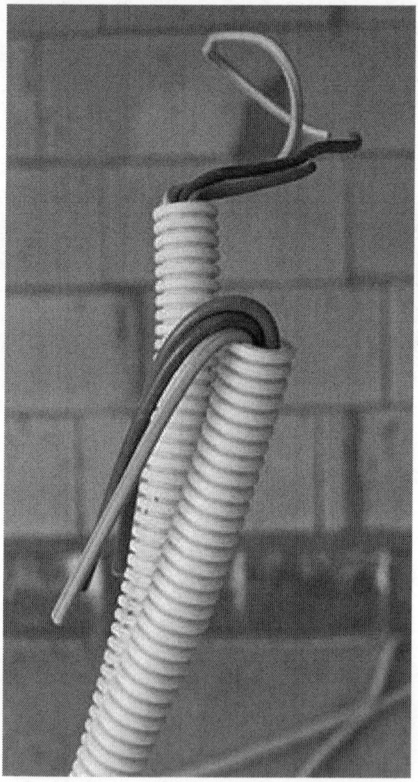

***Figure:*** *An electrical "3G" power cable found commonly in modern European houses. The cable consists of 3 wires (2 wires + 1 grounding in case if cable has "3G" name) and is double-insulated.*

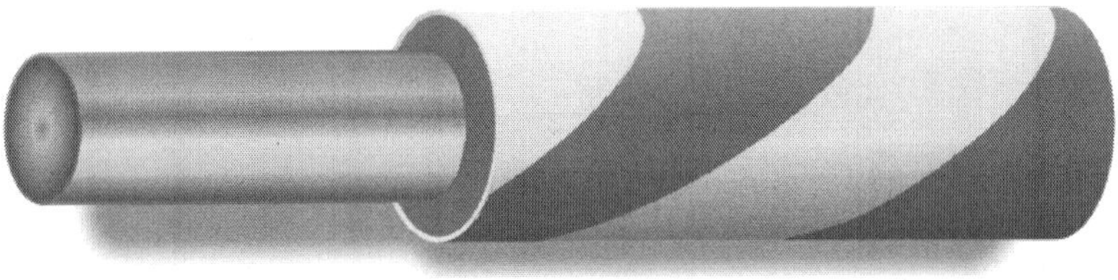

***Figure:*** *Diagram of typical green/yellow colour-coded wire for grounding (earthing).*

To enable wires to be easily and safely identified, all common wiring safety codes mandate a colour scheme for the insulation on power conductors. In a typical electrical code, some colour-coding is mandatory, while some may be optional. Many local rules and exceptions exist. Older installations vary in colour codes, and colours may shift with insulation exposure to heat, light, and aging.

Many electrical codes now recognise (or even require) the use of wire covered with green insulation, additionally marked with a prominent yellow stripe, for safety grounding (earthing) connections. This growing international standard was adopted for its distinctive appearance, to reduce the likelihood of dangerous confusion of safety grounding wires with other electrical functions, especially by persons affected by red-green colour blindness.

### Wiring Methods

Materials for wiring interior electrical systems in buildings vary depending on:

- Intended use and amount of power demand on the circuit
- Type of occupancy and size of the building
- National and local regulations
- Environment in which the wiring must operate.

***Figure:*** *Installing electrical wiring by cutting into the bricks of the building*

Wiring systems in a single family home or duplex, for example, are simple, with relatively low power requirements, infrequent changes to the building structure and layout, usually with dry, moderate temperature, and non-corrosive environmental conditions. In a light commercial environment, more frequent wiring changes can be expected, large apparatus may be installed, and special conditions of heat or moisture may apply. Heavy industries have more demanding wiring requirements, such as very large currents and higher voltages, frequent changes of equipment layout, corrosive, or wet or explosive atmospheres. In facilities that handle flammable gases or liquids, special rules may govern the installation and wiring of electrical equipment in hazardous areas. Wires and cables are rated by the circuit voltage, temperature rating, and environmental conditions (moisture, sunlight, oil, chemicals) in which they can be used. A wire or cable has a voltage (to neutral) rating, and a maximum conductor surface temperature rating. The amount of current a cable or wire can safely carry depends on the installation conditions.

### *Early Wiring Methods*

The first interior power wiring systems used conductors that were bare or covered with cloth, which were secured by staples to the framing of the building or on running boards. Where conductors went through walls, they were protected with cloth tape. Splices were done similarly to telegraph connections, and soldered for security. Underground conductors were insulated with wrappings of cloth tape soaked in pitch, and laid in wooden troughs which were then buried. Such wiring systems were unsatisfactory because of the danger of electrocution and fire, plus the high labour cost for such installations.

### *Knob and Tube*

The earliest standardized method of wiring in buildings, in common use in North America from about 1880 to the 1930s, was *knob and tube* (K&T) wiring: single conductors were run through cavities between the structural members in walls and ceilings, with ceramic tubes forming protective channels through joists and ceramic knobs attached to the structural members to provide air between the wire and the lumber and to support the wires. Since air was free to circulate over the wires, smaller conductors could be used than required in cables. By arranging wires on opposite sides of building structural members, some protection was afforded against short-circuits that can be caused by driving a nail into both conductors simultaneously.

By the 1940s, the labour cost of installing two conductors rather than one cable resulted in a decline in new knob-and-tube installations. However, the US code still allows new K&T wiring installations in special situations (some rural and industrial applications).

### *Metal-Sheathed Wires*

In the United Kingdom, an early form of insulated cable, introduced in 1896, consisted of two impregnated-paper-insulated conductors in an overall lead sheath. Joints were soldered, and special fittings were used for lamp holders and switches. These cables were similar to underground telegraph and telephone cables of the time. Paper-insulated cables proved unsuitable for interior wiring installations because very careful workmanship was required on the lead sheaths to ensure moisture did not affect the insulation.

A system later invented in the UK in 1908 employed vulcanized-rubber insulated wire enclosed in a strip metal sheath. The metal sheath was bonded to each metal wiring device to ensure earthing continuity.

A system developed in Germany called "Kuhlo wire" used one, two, or three rubber-insulated wires in a brass or lead-coated iron sheet tube, with a crimped seam. The enclosure could also be used as a return conductor. Kuhlo wire could be run exposed on surfaces and painted, or embedded in plaster. Special outlet and junction boxes were made for lamps and switches, made either of porcelain or sheet steel. The crimped seam was not considered as watertight as the *Stannos* wire used in England, which had a soldered sheath.

A somewhat similar system called "concentric wiring" was introduced in the United States around 1905. In this system, an insulated electrical wire was wrapped with copper tape which was then soldered, forming the grounded (return) conductor of the wiring system. The bare metal sheath, at earth potential, was considered safe to touch.

While companies such as General Electric manufactured fittings for the system, and a few buildings were wired with it, it was never adopted into the US National Electrical Code. Drawbacks of the system were that special fittings were required, and that any defect in the connection of the sheath would result in the sheath becoming energized.

### Other Historical Wiring Methods

Other methods of securing wiring that are now obsolete include:

- Re-use of existing gas pipes when converting gas light installations to electric lighting. Insulated conductors were pulled through the pipes that had formerly supplied the gas lamps. Although used occasionally, this method risked insulation damage from sharp edges inside the pipe at each joint.
- Wood mouldings with grooves cut for single conductor wires, covered by a wooden cap strip. These were prohibited in North American electrical codes by 1928. Wooden moulding was also used to some degree in England, but was never permitted by German and Austrian rules.

- A system of flexible twin cords supported by glass or porcelain buttons was used near the turn of the 20th century in Europe, but was soon replaced by other methods.
- During the first years of the 20th century, various patented forms of wiring system such as Bergman and Peschel tubing were used to protect wiring; these used very thin fibre tubes, or metal tubes which were also used as return conductors.
- In Austria, wires were concealed by embedding a rubber tube in a groove in the wall, plastering over it, then removing the tube and pulling wires through the cavity.

Metal moulding systems, with a flattened oval section consisting of a base strip and a snap-on cap channel, were more costly than open wiring or wooden moulding, but could be easily run on wall surfaces. Similar surface mounted raceway wiring systems are still available today.

### Cables

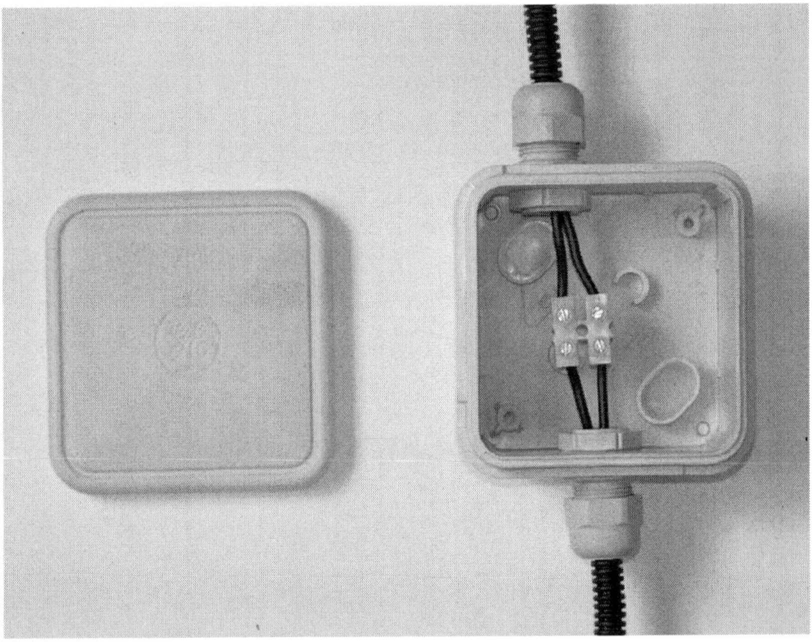

***Figure:*** *Wiring for extremely wet conditions*

Armoured cables with two rubber-insulated conductors in a flexible metal sheath were used as early as 1906, and were considered at the time a better method than open knob-and-tube wiring, although much more expensive.

The first polymer-insulated cables for building wiring were introduced in 1922. These were two or more solid copper electrical wires with rubber insulation, plus woven cotton cloth over each conductor for protection of the insulation, with an overall woven jacket, usually impregnated with tar as a protection from moisture. Waxed paper was used as a filler and separator.

Over time, rubber-insulated cables become brittle because of exposure to atmospheric oxygen, so they must be handled with care, and are usually replaced during renovations. When switches, outlets or light fixtures are replaced, the mere act of tightening connections may cause hardened insulation to flake off the conductors. Rubber insulation further inside the cable often is in better condition than the insulation exposed at connections, due to reduced exposure to oxygen.

Rubber insulation was hard to strip from bare copper, so copper was tinned, causing slightly more electrical resistance. Rubber insulation is no longer used for permanent wiring installations, but may still be used for replaceable temporary cables where flexibility is important, such as electrical extension cords.

About 1950, PVC insulation and jackets were introduced, especially for residential wiring. About the same time, single conductors with a thinner PVC insulation and a thin nylon jacket (e.g. US Type THN, THHN, etc.) became common.

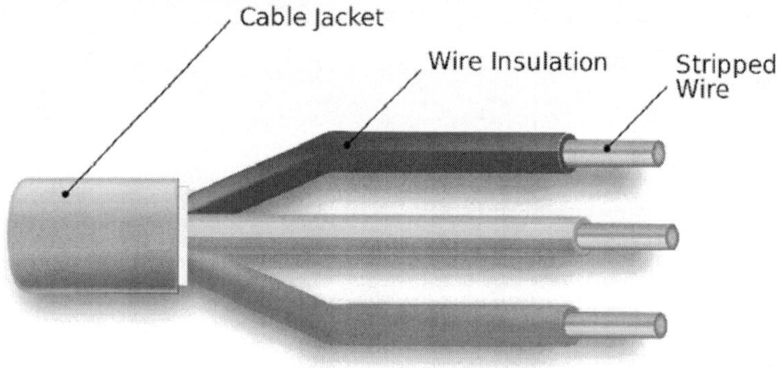


***Figure:*** *Diagram of a simple electrical cable with three insulated conductors*

The simplest form of cable has two insulated conductors twisted together to form a unit; such unjacketed cables with two or three conductors are used for low-voltage signal and control applications such as doorbell wiring. In North American practice, an overhead cable from a transformer on a power pole to a residential electrical service consists of three twisted (triplexed) wires, often with one being a bare wire made of copper (protective earth/ground) and the other two being insulated for the line voltage (hot/live wire and neutral wire). For additional safety, the ground wire may be formed into a stranded co-axial layer completely surrounding the phase conductors, so that the outmost conductor is grounded.

### Copper Conductors

Electrical devices often contain copper conductors because of their multiple beneficial properties, including their high electrical conductivity, tensile strength, ductility, creep resistance, corrosion resistance, thermal conductivity, coefficient of thermal expansion, solderability, resistance to electrical overloads, compatibility with electrical insulators, and ease of installation.

Despite competition from other materials, copper remains the preferred electrical conductor in nearly all categories of electrical wiring. For example, copper is used to conduct electricity in high, medium and low voltage power networks, including power generation, power transmission, power distribution, telecommunications, electronics circuitry, data processing, instrumentation, appliances, entertainment systems, motors, transformers, heavy industrial machinery, and countless other types of electrical equipment.

### Aluminium Conductors

Aluminium wire was common in North American residential wiring from the late 1960s to mid-1970s due to the rising cost of copper. Because of its greater resistivity, aluminium wiring requires larger conductors than copper. For instance, instead of 14 AWG (American wire gauge) for most lighting circuits, aluminium wiring would be 12 AWG on a typical 15 ampere circuit, though local building codes may vary. Aluminium conductors were originally indiscriminately used with wiring devices intended for copper conductors. This practice was found to cause defective connections unless the aluminium was one of a special alloy, or all devices — breakers, switches, receptacles, splice connectors, wire nuts, etc. — were specially designed for the purpose. These special designs address problems with junctions between dissimilar metals, oxidation on metal surfaces, and mechanical effects that occur as different metals expand at different rates with increases in temperature.

Unlike copper, aluminium has a tendency to cold-flow under pressure, so screw clamped connections may get loose over time. This can be mitigated by using spring-loaded connectors that apply constant pressure, applying high pressure cold joints in splices and termination fittings, and torquing the bolted connection.

Also unlike copper, aluminium forms an insulating oxide layer on the surface. This is sometimes addressed by coating aluminium wires with an antioxidant paste at joints, or by applying a mechanical termination designed to break through the oxide layer during installation. Because of improper design and installation, some junctions to wiring devices would overheat under heavy current load, and cause fires. Revised standards for wiring devices (such as the CO/ ALR "copper-aluminium-revised" designation) were developed to reduce these problems. Nonetheless, aluminium wiring for residential use has acquired a poor reputation and has fallen out of favour.

Aluminium conductors are still used for bulk power distribution and large feeder circuits, because they cost less than copper wiring, and weigh less, especially in the large sizes needed for heavy current loads. Aluminium conductors must be installed with compatible connectors.

### Modern Wiring Materials

Modern non-metallic sheathed cables, such as (US and Canadian) Type NMB and NMC, consist of two to four wires covered with thermoplastic insulation and a bare wire for grounding (bonding) surrounded by a flexible plastic jacket. Some versions wrap the individual conductors in paper before the plastic jacket is applied. Romex brand NM cable is the most well-known brand of NM cable, since the first of its type was manufactured by Rome Cable Division of Cyprus Mines, Rome, New York. The Romex trademark has been owned by Southwire Company since it purchased the electrical building wire assets of General Cable in 2001.

Special versions of non-metallic sheathed cables, such as US Type UF, are designed for direct underground burial (often with separate mechanical protection) or exterior use where exposure to ultraviolet radiation (UV) is a possibility. These cables differ in having a moisture-resistant construction, lacking paper or other absorbent fillers, and being formulated for UV resistance.

Rubber-like synthetic polymer insulation is used in industrial cables and power cables installed underground because of its superior moisture resistance. Insulated cables are rated by their allowable operating voltage and their maximum operating temperature at the conductor surface.

A cable may carry multiple usage ratings for applications, for example, one rating for dry installations and another when exposed to moisture or oil.

Generally, single conductor building wire in small sizes is solid wire, since the wiring is not required to be very flexible. Building wire conductors larger than 10 AWG (or about 6 mm$^2$) are stranded for flexibility during installation, but not stranded enough to be flexible enough to use as appliance cord.

Cables for industrial, commercial, and apartment buildings may contain many insulated conductors in an overall jacket, with helical tape steel or aluminium armour, or steel wire armour, and perhaps as well an overall PVC or lead jacket for protection from moisture and physical damage. Cables intended for very flexible service or in marine applications may be protected by woven bronze wires. Power or communications cables (e.g., computer networking) that are routed in or through air-handling spaces (plenums) of office buildings are required under the model code to be either encased in metal conduit or rated for low flame and smoke production.

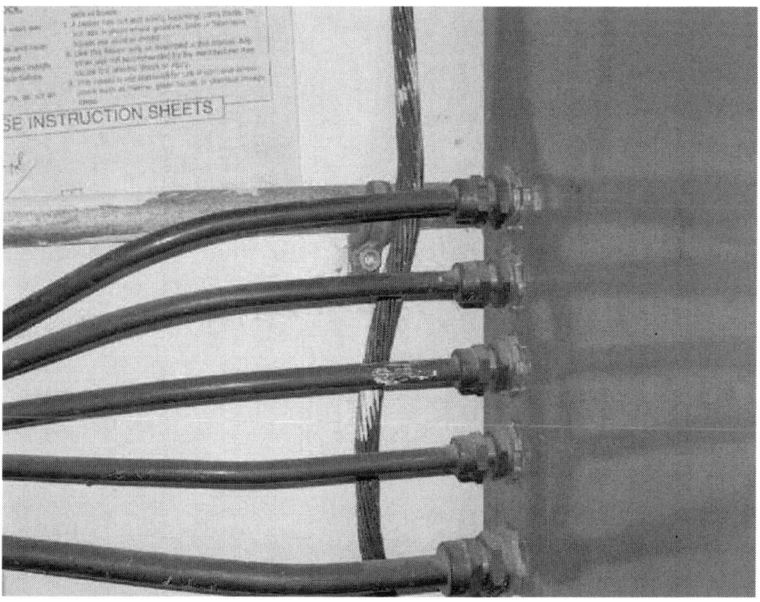


***Figure:*** *Mineral insulated cables at a panel board*

For some industrial uses in steel mills and similar hot environments, no organic material gives satisfactory service. Cables insulated with compressed mica flakes are sometimes used. Another form of high-temperature cable is a mineral insulated cable, with individual conductors placed within a copper tube, and the space filled with magnesium oxide powder. The whole assembly is drawn down to smaller sizes, thereby compressing the powder. Such cables have a certified fire resistance rating, are more costly than non-fire rated cable, and have little flexibility and are effectively rigid to the user of the cable.

Because multiple conductors bundled in a cable cannot dissipate heat as easily as single insulated conductors, those circuits are always rated at a lower "ampacity". Tables in electrical safety codes give the maximum allowable current for a particular size of conductor, for the voltage and temperature rating at the surface of the conductor for a given physical environment, including the insulation type and thickness. The allowable current will be different for wet or dry, for hot (attic) or cool (underground) locations. In a run of cable through several areas, the most severe area will determine the appropriate rating of the overall run.

Cables usually are secured by special fittings where they enter electrical apparatus; this may be a simple screw clamp for jacketed cables in a dry location, or a polymer-gasketed cable connector that mechanically engages the armour of an armoured cable and provides a water-resistant connection. Special cable fittings may be applied to prevent explosive gases from flowing in the interior of jacketed cables, where the cable passes through areas where inflammable gases are present. To prevent loosening of the connections of individual conductors of a cable, cables must be supported near their entrance to devices and at regular intervals through their length. In tall buildings special designs are required to support the conductors of vertical runs of cable. Usually, only one cable per fitting is allowed unless the fitting is otherwise rated.

Special cable constructions and termination techniques are required for cables installed in ocean-going vessels; in addition to electrical safety and fire safety, such cables may also be required to be pressure-resistant where they penetrate bulkheads of a ship.

### Raceways

Insulated wires may be run in one of several forms of a raceway between electrical devices. This may be a specialised bendable pipe, called a conduit, or one of several varieties of metal (rigid steel or aluminium) or non-metallic (PVC or HDPE) tubing. Rectangular cross-section metal or PVC wire troughs (North America) or trunking (UK) may be used if many circuits are required. Wires run underground may be run in plastic tubing encased in concrete, but metal elbows may be used in severe pulls. Wiring in exposed areas, for example factory floors, may be run in cable trays or rectangular raceways having lids.

Where wiring, or raceways that hold the wiring, must traverse fire-resistance rated walls and floors, the openings are required by local building codes to be firestopped. In cases where safety-critical wiring must be kept operational during an accidental fire, fireproofing must be applied to maintain circuit integrity in a manner to comply with a product's certification listing. The nature and thickness of any passive fire protection materials used in conjunction with wiring and raceways has a quantifiable impact upon the ampacityderating, because the thermal insulation properties needed for fire resistance also inhibit air cooling of power conductors.

Cable trays are used in industrial areas where many insulated cables are run together. Individual cables can exit the tray at any point, simplifying the wiring installation and reducing the labour cost for installing new cables. Power cables may have fittings in the tray to maintain clearance between the conductors, but small control wiring is often installed without any intentional spacing between cables.

Since wires run in conduits or underground cannot dissipate heat as easily as in open air, and since adjacent circuits contribute induced currents, wiring regulations give rules to establish the current capacity (ampacity). Special sealed fittings are used for wiring routed through potentially explosive atmospheres.

### Bus Bars, Bus Duct, Cable Bus

For very heavy currents in electrical apparatus, and for heavy currents distributed through a building, bus bars can be used. (The term "bus" is a contraction of the Latin *omnibus* - meaning "for all".) Each live conductor of such a system is a rigid piece of copper or aluminium, usually in flat bars (but sometimes as tubing or other shapes). Open bus bars are never used in publicly accessible areas, although they are used in manufacturing plants and power company switch yards to gain the benefit of air cooling. A variation is to use heavy cables, especially where it is desirable to transpose or "roll" phases.

In industrial applications, conductor bars are often pre-assembled with insulators in grounded enclosures. This assembly, known as bus duct or busway, can be used for connections to large switchgear or for bringing the main power feed into a building. A form of bus duct known as "plug-in bus" is used to distribute power down the length of a building; it is constructed to allow tap-off switches or motor controllers to be installed at designated places along the bus. The big advantage of this scheme is the ability to remove or add a branch circuit without removing voltage from the whole duct.

Bus ducts may have all phase conductors in the same enclosure (non-isolated bus), or may have each conductor separated by a grounded barrier from the adjacent phases (segregated bus). For conducting large currents between devices, a cable bus is used.

For very large currents in generating stations or substations, where it is difficult to provide circuit protection, an isolated-phase bus is used. Each phase of the circuit is run in a separate grounded metal enclosure. The only fault possible is a phase-to-ground fault, since the enclosures are separated. This type of bus can be rated up to 50,000 amperes and up to hundreds of kilovolts (during normal service, not just for faults), but is not used for building wiring in the conventional sense.

### *Electrical Panels*

Electrical panels are easily accessible junction boxes used to reroute and switch electrical services. The term is often used to refer to circuit breaker panels or "fuseboxes".

3

# Construction Model

## Elevator

The elevator (or lift in the Commonwealth excluding Canada) is a type of vertical transport equipment that efficiently moves people or goods between floors (levels, decks) of a building, vessel or other structure. Elevators are generally powered by electric motors that either drive traction cables or counterweight systems like a hoist, or pump hydraulic fluid to raise a cylindrical piston like a jack.

***Figure:** Elevator lobby at the Forest Glen Washington Metro station in Silver Spring, Maryland.*

In agriculture and manufacturing, an elevator is any type of conveyor device used to lift materials in a continuous stream into bins or silos. Several types exist, such as the chain and bucket elevator, grain auger screw conveyor using the principle of Archimedes' screw, or the chain and paddles/forks of hay elevators.

Languages other than English may have loanwords based on either *elevator* (e.g., Korean and Japanese) or *lift* (e.g., Russian and Cantonese). Because of wheelchair access laws, elevators are often a legal requirement in new multistory buildings, especially where wheelchair ramps would be impractical.

## History

***Pre-industrial era:*** The first reference to an elevator is in the works of the Roman architect Vitruvius, who reported that Archimedes (c. 287 BC – c. 212 BC) built his first elevator probably in 236 BC. In some literary sources of later historical periods, elevators were mentioned as cabs on a hemp rope and powered by hand or by animals. It is supposed that elevators of this type were installed in the Sinai monastery of Egypt.

In 1000, the *Book of Secrets* by al-Muradi in Islamic Spain described the use of an elevator-like lifting device, in order to raise a large battering ram to destroy a fortress. In the 17th century the prototypes of elevators were located in the palace buildings of England and France.

Ancient and medieval elevators used drive systems based on hoists or winders. The invention of a system based on the screw drive was perhaps the most important step in elevator technology since ancient times, leading to the creation of modern passenger elevators. The first screw drive elevator was built by Ivan Kulibin and installed in Winter Palace in 1793. Several years later another of Kulibin's elevators was installed in Arkhangelskoye near Moscow.

## Industrial Era

The development of elevators was led by the need for movement of raw materials including coal and lumber from hillsides. The technology developed by these industries and the introduction of steel beam construction worked together to provide the passenger and freight elevators in use today.

Starting in the coal mines, by the mid-19th century elevators were operated with steam power and were used for moving goods in bulk in mines and factories. These steam driven devices were soon being applied to a diverse set of purposes - in 1823, two architects working in London, Burton and Hormer, built and operated a novel tourist attraction, which they called the "ascending room". It elevated paying customers to a considerable height in the centre of London, allowing them a magnificent panoramic view of the city centre.

Early, crude steam-driven elevators were refined in the ensuing decade; - in 1835 an innovative elevator called the "Teagle" was developed by the company Frost and Stutt in England. The elevator was belt-driven and used a counterweight for extra power.

The hydraulic crane was invented by Sir William Armstrong in 1846, primarily for use at the Tyneside docks for loading cargo. These quickly supplanted the earlier steam driven lifts as they were able to leverage Pascal's law to provide a much greater force. A water pump supplied a variable level of water pressure to a plunger encased inside a vertical cylinder, allowing the level of the platform (carrying a heavy load) to be raised and lowered. Counter - weights and balances were also used to increase the lifting power of the apparatus.

Henry Waterman of New York is credited with inventing the "standing rope control" for an elevator in 1850.

In 1852, Elisha Otis introduced the safety elevator, which prevented the fall of the cab if the cable broke. The design of the Otis safety elevator is somewhat similar to one type still used today. A governor device engages knurled roller(s), locking the elevator to its guides should the elevator descend at excessive speed. He demonstrated it at the New York exposition in the Crystal Palace in a dramatic, death-defying presentation in 1854, and the first such passenger elevator was installed at 488 Broadway in New York City on March 23, 1857.

The first elevator shaft preceded the first elevator by four years. Construction for Peter Cooper's Cooper Union Foundation building in New York began in 1853. An elevator shaft was included in the design, because Cooper was confident that a safe passenger elevator would soon be invented. The shaft was cylindrical because Cooper felt it was the most efficient design. Later, Otis designed a special elevator for the building. Today the Otis Elevator Company, now a subsidiary of United Technologies Corporation, is the world's largest manufacturer of vertical transport systems.

The Equitable Life Building completed in 1870 in New York City was the first office building to have passenger elevators. The first electric elevator was built by Werner von Siemens in 1880 in Germany. The inventor Anton Freissler developed the ideas of von Siemens and built up a successful enterprise in Austria-Hungary. The safety and speed of electric elevators were significantly enhanced by Frank Sprague who added floor control, automatic elevators, acceleration control of cars, and safeties. His elevator ran faster and with larger loads than hydraulic or steam elevators, and 584 electric elevators were installed before Sprague sold his company to the Otis Elevator Company in 1895. Sprague also developed the idea and technology for multiple elevators in a single shaft.

In 1882, when hydraulic power was a well established technology, a company later named the London Hydraulic Power Company was formed. It constructed a network of high-pressure mains on both sides of the Thames which, ultimately, extended to 184 miles and powered some 8,000 machines, predominantly lifts (elevators) and cranes.

In 1874, J.W. Meaker patented a method which permitted elevator doors to open and close safely. In 1887, American Inventor Alexander Miles of Duluth, Minnesota patented an elevator with automatic doors that would close off the elevator shaft.

In 2000 a vacuum elevator was offered commercially in Argentina.

### Design

Some people argue that elevators began as simple rope or chainhoists. An elevator is essentially a platform that is either pulled or pushed up by a mechanical means. A modern day elevator consists of a cab (also called a "cage" or "car") mounted on a platform within an enclosed space called a shaft or sometimes a "hoistway". In the past, elevator drive mechanisms were powered by steam and water hydraulic pistons or by hand. In a "traction" elevator, cars are pulled up by means of rolling steel ropes over a deeply grooved pulley, commonly called a sheave in the industry. The weight of the car is balanced by a counterweight. Sometimes two elevators are built so that their cars always move synchronously in opposite directions, and are each other's counterweight.

The friction between the ropes and the pulley furnishes the traction which gives this type of lift its name.

Hydraulic elevators use the principles of hydraulics (in the sense of hydraulic power) to pressurize an above ground or in-ground piston to raise and lower the car. Roped hydraulics use a combination of both ropes and hydraulic power to raise and lower cars. Recent innovations include permanent magnet motors, machine room-less rail mounted gearless machines, and microprocessor controls.

The technology used in new installations depends on a variety of factors. Hydraulic elevators are cheaper, but installing cylinders greater than a certain length becomes impractical for very-high lift hoistways. For buildings of much over seven storys, traction lifts must be employed instead. Hydraulic elevators are usually slower than traction lifts.

Elevators are a candidate for mass customization. There are economies to be made from mass production of the components, but each building comes with its own requirements like different number of floors, dimensions of the well and usage patterns.

### *Elevator Doors*

Elevator doors protect riders from falling into the shaft. The most common configuration is to have two panels that meet in the middle, and slide open laterally. In a cascading telescopic configuration (potentially allowing wider entryways within limited space), the doors run on independent tracks so that while open, they are tucked behind one another, and while closed, they form cascading layers on one side.

This can be configured so that two sets of such cascading doors operate like the centre opening doors described above, allowing for a very wide elevator cab. In less expensive installations the elevator can also use one large "slab" door: a single panel door the width of the doorway that opens to the left or right laterally. Some buildings have elevators with the single door on the shaft way, and double cascading doors on the cab.

### *Machine Room-Less (MRL) Elevators*

Machine room-less elevators are designed so that most of the components fit within the shaft containing the elevator car; and a small cabinet houses the elevator controller. Other than the machinery being in the hoistway, the equipment is similar to a normal traction elevator. The benefits are:

- creates more usable space
- use less energy (70-80% less than hydraulic elevators)
- uses no oil
- all components are above ground similar to roped hydraulic type elevators (this takes away the environmental concern that was created by the hydraulic cylinder on direct hydraulic type elevators being stored underground)
- slightly lower cost than other elevators

- can operate at faster speeds than hydraulics but not normal traction units.

Detriments

- Equipment can be harder to service and maintain.
- No code has been approved for the installation of residential elevator equipment.

Facts

- Noise level is at 50-55 dBA (A-weighted decibels), which can be lower than some but not all types of elevators.
- Usually used for low-rise to mid-rise buildings
- The motor mechanism is placed in the hoistway itself
- The US was slow to accept the commercial MRL Elevator because of codes

—national and local building codes did not address elevators without machine rooms. Residential MRL Elevators are still not allowed by the ASME A17 code in the US. MRL elevators have been recognised in the 2005 supplement to the 2004 A17.1 Elevator Code.

***Figure:*** *KoneEcoDisc, Notice how the entire drive system is in the hoistway.*

### Elevator Traffic Calculations

***Round Trip Time Calculations:*** The majority of elevator designs are developed from Up Peak Round Trip Time calculations as described in the following publications:- CIBSE Guide D: Transportation Systems in Building Elevator Traffic Handbook, Theory and Practice. Gina Barney The Vertical Transportation Handbook. George Strakosch

Traditionally, these calculations have formed the basis of establishing the Handling Capacity of an elevator system. Modern Installations with more complex elevator arrangements have led to the development of more specific formulae such as the General Analysis calculation.

Subsequently this has been extended for Double Deck elevators.

### *Simulation*

Elevator traffic simulation software can be used to model complex traffic patterns and elevator arrangements that cannot necessarily be analysed by RTT calculations.

### *Lift Traffic Patterns*

There are four main types of elevator traffic patterns that can be observed in most modern office installations. They are up peak traffic, down peak traffic, lunch time (two way) traffic and interfloor traffic.

### *Types of Hoist Mechanisms*

There are at least four means of moving an elevator:

### *Traction Elevators*

- *Geared and gearless traction elevators*

Geared traction machines are driven by AC or DC electric motors. Geared machines use worm gears to control mechanical movement of elevator cars by "rolling" steel hoist ropes over a drive sheave which is attached to a gearbox driven by a high-speed motor. These machines are generally the best option for basement or overhead traction use for speeds up to 500 feet per minute (3 m/s). In order to allow accurate speed control of the motor, to allow accurate levelling and for passenger comfort, a DC hoist motor powered by an AC/DC motor-generator (MG) set was the preferred solution in high-traffic elevator installations for many decades.

The MG set also typically powered the relay controller of the elevator, which has the added advantage of electrically isolating the elevators from the rest of a building's electrical system, thus eliminating the transient power spikes in the building's electrical supply caused by the motors starting and stopping (causing lighting to dim every time the elevators are used for example), as well as interference to other electrical equipment caused by the arcing of the relay contactors in the control system.

Contemporary cheaper installations, such as those in residential buildings and low-traffic commercial applications generally used a single or two-speed AC hoist machine. The widespread availability of cheap solid state AC drives has allowed infinitely variable speed AC motors to be used universally, bringing with it the advantages of the older motor-generator based systems, without the penalties in terms of efficiency and complexity. The older MG-based installations are gradually being replaced in older buildings due to their poor energy efficiency.

Gearless traction machines are low-speed (low-RPM), high-torque electric motors powered either by AC or DC. In this case, the drive sheave is directly attached to the end of the motor. Gearless traction elevators can reach speeds of up to 2,000 feet per minute (10 m/s), or even higher. A brake is mounted between the motor and drive sheave (or gearbox) to hold the elevator stationary at a floor. This brake is usually an external drum type and is actuated by spring force and held open electrically; a power failure will cause the brake to engage and prevent the elevator from falling.

In each case, cables are attached to a hitch plate on top of the cab or may be "underslung" below a cab, and then looped over the drive sheave to a counterweight attached to the opposite end of the cables which reduces the amount of power needed to move the cab. The counterweight is located in the hoist-way and rides a separate railway system; as the car goes up, the counterweight goes down, and vice versa. This action is powered by the traction machine which is directed by the controller, typically a relay logic or computerized device that directs starting, acceleration, deceleration and stopping of the elevator cab. The weight of the counterweight is typically equal to the weight of the elevator cab plus 40-50% of the capacity of the elevator.

The grooves in the drive sheave are specially designed to prevent the cables from slipping. "Traction" is provided to the ropes by the grip of the grooves in the sheave, thereby the name. As the ropes age and the traction grooves wear, some traction is lost and the ropes must be replaced and the sheave repaired or replaced. Sheave and rope wear may be significantly reduced by ensuring that all ropes have equal tension, thus sharing the load evenly. Rope tension equalisation may be achieved using a rope tension gauge, and is a simple way to extend the lifetime of the sheaves and ropes.

Elevators with more than 100 feet (30 m) of travel have a system called compensation. This is a separate set of cables or a chain attached to the bottom of the counterweight and the bottom of the elevator cab. This makes it easier to control the elevator, as it compensates for the differing weight of cable between the hoist and the cab. If the elevator cab is at the top of the hoist-way, there is a short length of hoist cable above the car and a long length of compensating cable below the car and vice versa for the counterweight. If the compensation system uses cables, there will be an additional sheave in the pit below the elevator, to guide the cables. If the compensation system uses chains, the chain is guided by a bar mounted between the counterweight railway lines.

### Hydraulic Elevators

- *Conventional hydraulic elevators.* They use an underground cylinder, are quite common for low level buildings with 2–5 floors (sometimes but seldom up to 6–8 floors), and have speeds of up to 200 feet per minute (1 m/s).
- *Holeless hydraulic elevators* were developed in the 1970s, and use a pair of above ground cylinders, which makes it practical for environmentally or cost sensitive buildings with 2, 3, or 4 floors.
- *Roped hydraulic elevators* use both above ground cylinders and a rope system, allowing the elevator to travel further than the piston has to move.

The low mechanical complexity of hydraulic elevators in comparison to traction elevators makes them ideal for low rise, low traffic installations. They are less energy efficient as the pump works against gravity to push the car and its passengers upwards; this energy is lost when the car descends on its own weight. The high current draw of the pump when starting up also places higher demands on a building's electrical system. There are also environmental concerns should either the lifting cylinder leak fluid into the ground.

The modern generation of low cost, machine room-less traction elevators made possible by advances in miniaturization of the traction motor and control systems challenges the supremacy of the hydraulic elevator in their traditional market niche.

### Traction-Hydraulic Elevators

The traction-hydraulic elevator has overhead traction cables and counterweight, but is driven by hydraulic power instead of an overhead traction motor. The weight of the car and its passengers, plus an advantageous roping ratio, reduces the demand from the pump to raise the counterweight, thereby reducing the size of the required machinery.

### Climbing Elevator

A climbing elevator is a self-ascending elevator with its own propulsion. The propulsion can be done by an electric or a combustion engine. Climbing elevators are used in guyed masts or towers, in order to make easy access to parts of these constructions, such as flight safety lamps for maintenance. An example would be the Moonlight towers in Austin, Texas, where the elevator holds only one person and equipment for maintenance. The Glasgow Tower — an observation tower in Glasgow, Scotland — also makes use of two climbing elevators.

### Controlling Elevators

Early elevators had no automatic landing positioning. Elevators were operated by elevator operators using a motor controller. The controller was contained within a cylindrical container about the size and shape of a cake container and this was operated via a projecting handle. This allowed some control over the energy supplied to the motor (located at the top of the elevator shaft or beside the bottom of the elevator shaft) and so enabled the elevator to be accurately positioned — if the operator was sufficiently skilled. More typically the operator would have to "jog" the control to get the elevator reasonably close to the landing point and then direct the outgoing and incoming passengers to "watch the step". Some older freight elevators are controlled by switches operated by pulling on adjacent ropes. Safety interlocks ensure that the inner and outer doors are closed before the elevator is allowed to move. Most older manually controlled elevators have been retrofitted with automatic or semi-automatic controls.

Automatic elevators began to appear as early as the 1930s, their development being hastened by striking elevator operators which brought large cities dependent on skyscrapers (and therefore their elevators) such as New York and Chicago to their knees. These electromechanical systems used relay logic circuits of increasing complexity to control the speed, position and door operation of an elevator or bank of elevators. The Otis *Autotronic* system of the early 1950s brought the earliest predictive systems which could anticipate traffic patterns within a building to deploy elevator movement in the most efficient manner. Relay-controlled elevator systems remained common until the 1980s, and their gradual replacement with solid-state microprocessor based controls which are now the industry standard.

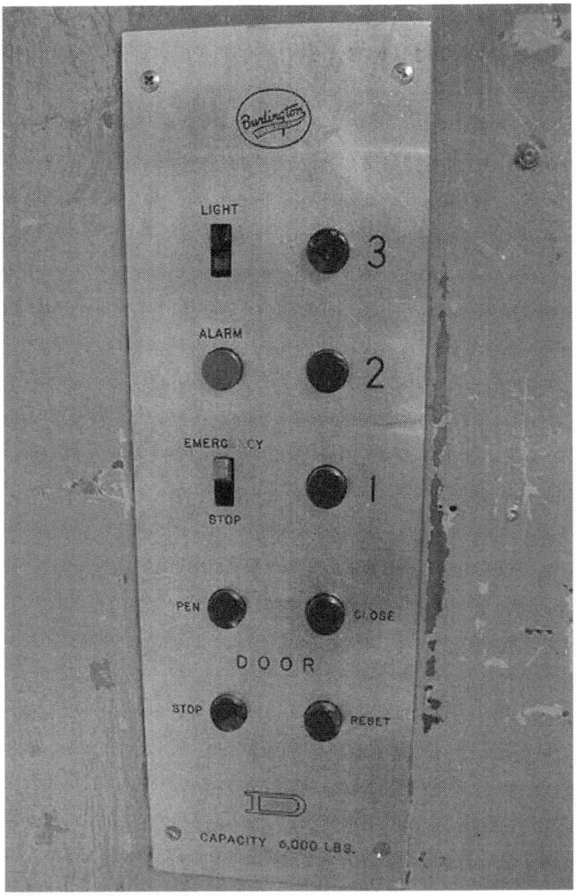


***Figure:*** *Typical freight elevator control station*

***Figure:*** *Typical passenger elevator control station*

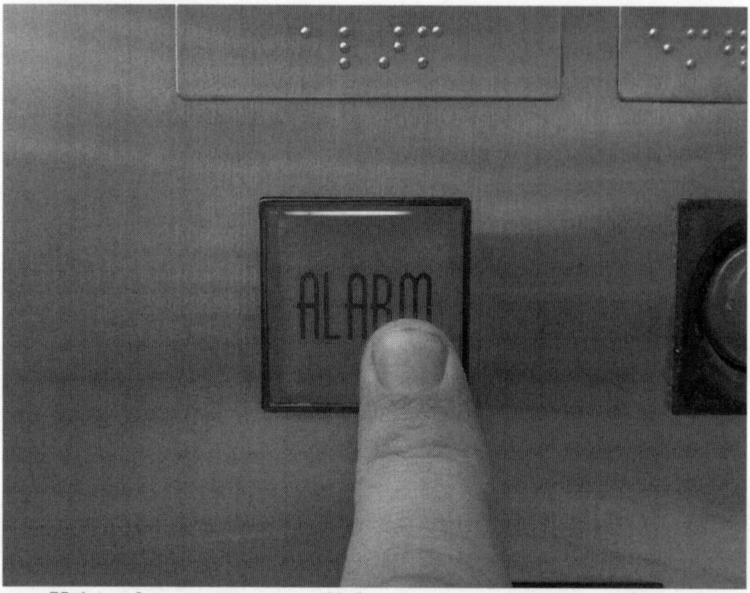


***Figure:*** *Using the emergency call button in an elevator. There is Braille text for visually impaired people and the button glows to alert a hearing impaired person that the bell is ringing and the call is being placed*

## General Controls

A typical modern passenger elevator will have:

- Space to stand in, guardrails, seating cushion (luxury)
- Overload sensor — prevents the elevator from moving until excess load has been removed. It may trigger a voice prompt or buzzer alarm. This may also trigger a "full car" indicator, indicating the car's inability to accept more passengers until some are unloaded.
- Electric fans or air conditioning units to enhance circulation and comfort.
- Call buttons to choose a floor. Some of these may be key switches (to control access). In some elevators, certain floors are inaccessible unless one swipes a security card or enters a passcode (or both). In the United States and other countries, call button text and icons are raised to allow blind users to operate the elevator; many have Braille text besides.
- A set of doors kept locked on each floor to prevent unintentional access into the elevator shaft by the unsuspecting individual. The door is unlocked and opened by a machine sitting on the roof of the car, which also drives the doors that travel with

the car. Door controls are provided to close immediately or reopen the doors, although the button to close them immediately is often disabled during normal operations, especially on more recent elevators. Objects in the path of the moving doors will either be detected by sensors or physically activate a switch that reopens the doors. Otherwise, the doors will close after a preset time.

- A stop switch (not allowed under British regulations) to halt the elevator while in motion and often used to hold an elevator open while freight is loaded. Keeping an elevator stopped for too long may trigger an alarm. Unless local codes require otherwise, this will most likely be a key switch.
- An alarm button or switch, which passengers can use to signal that they have been trapped in the elevator.

Some elevators may have one or more of the following:

- An elevator telephone, which can be used (in addition to the alarm) by a trapped passenger to call for help.
- Hold button: This button delays the door closing timer, useful for loading freight and hospital beds.
- Call cancellation: A destination floor may be deselected by double clicking.
- Access restriction by key switches, RFID reader, code keypad, hotel room card, etc..
- One or more additional sets of doors that can serve different floor plans. For example, in an elevated crosswalk setup, the front doors may open on the street level, and the rear doors open on the crosswalk level.
- Security camera
- Plain walls or mirrored walls.
- Glass windowpane providing a view of the building interior or onto the streets.

Other controls, which are generally inaccessible to the public (either because they are key switches, or because they are kept behind a locked panel), include:

- Fireman's service, phase II key switch
- Switch to enable or disable the elevator.
- An *inspector's* switch, which places the elevator in inspection mode (this may be situated on top of the elevator)

- Manual up/down controls for elevator technicians, to be used in inspection mode, for example.
- An *independent service/exclusive mode* (also known as "Car Preference"), which will prevent the car from answering to hall calls and only arrive at floors selected via the panel.

  The door should stay open while parked on a floor. This mode may be used for temporarily transporting goods.
- Attendant service mode.

***Figure:*** *Manual pushbutton elevator controls.*

- Large buildings with multiple elevators of this type also had an *elevator dispatcher* stationed in the lobby to direct passengers and to signal the operator to leave with the use of a mechanical "cricket" noisemaker.

### External Controls

Elevators are typically controlled from the outside by up and down buttons at each stop. When pressed at a certain floor, the elevator arrives to pick up more passengers. If the particular elevator is currently serving traffic in a certain direction, it will only answer hall calls in the same direction unless there are no more calls beyond that floor.

***Figure:*** *An external control panel*

In a group of two or more elevators, the call buttons may be linked to a central dispatch computer, such that they illuminate and cancel together. This is done to ensure that only one car is called at one time.

Key switches may be installed on the ground floor so that the elevator can be remotely switched on or off from the outside.

In destination control systems, one selects the intended destination floor (in lieu of pressing *"up"* or *"down"*) and is then notified which elevator will serve their request.

### The Elevator Algorithm

The elevator algorithm, a simple algorithm by which a single elevator can decide where to stop, is summarized as follows:

- Continue travelling in the same direction while there are remaining requests in that same direction.
- If there are no further requests in that direction, then stop and become idle, or change direction if there are requests in the opposite direction.

The elevator algorithm has found an application in computer operating systems as an algorithm for scheduling hard disk requests. Modern elevators use more complex heuristic algorithms to decide which request to service next. An introduction to these algorithms can be found in the "Elevator traffic handbook: theory and practice" given in the references below.

### Destination Control System

Some skyscraper buildings and other types of installation feature a destination operating panel where a passenger registers their floor calls before entering the car. The system lets them know which car to wait for, instead of everyone boarding the next car. In this way, travel time is reduced as the elevator makes fewer stops for individual passengers, and the computer distributes adjacent stops to different cars in the bank. Although travel time is reduced passenger waiting times may be longer as they will not necessarily be allocated the next car to depart. During the down peak period the benefit of destination control will be limited as passengers have a common destination.

It can also improve accessibility, as a mobility-impaired passenger can move to his or her designated car in advance.

Inside the elevator there is no call button to push, or the buttons are there but they cannot be pushed — except door opening and alarm button — they only indicate stopping floors.

The idea of destination control was originally conceived by Leo Port from Sydney in 1961 but at that time lift controllers were implemented in relays and were unable to optimise the performance of destination control allocations.

The system was first pioneered by Schindler Elevator in 1992 as the Miconic 10. Manufacturers of such systems claim that average travelling time can be reduced by up to 30%.

However, performance enhancements cannot be generalized as the benefits and limitations of the system are dependent on many factors. One problem is that the system is subject to gaming. Sometimes, one person enters the destination for a large group of people going to the same floor. The dispatching algorithm is usually unable to completely cater for the variation, and latecomers may find the elevator they are assigned to is already full. Also, occasionally, one person may press the floor multiple times. This is common with up/down buttons when people believe this to be an effective way to hurry elevators. However, this will make the computer think multiple people are waiting and will allocate empty cars to serve this one person.

To prevent this problem, in one implementation of destination control, every user gets an RFID card to identify himself so the system knows every user call and can cancel the first call if the passenger decides to travel to another destination to prevent empty calls. The newest invention knows even where people are located and how many on which floor because of their identification, either for the purposes of evacuating the building or for security reasons.

The same destination scheduling concept can also be applied to public transit such as in group rapid transit.

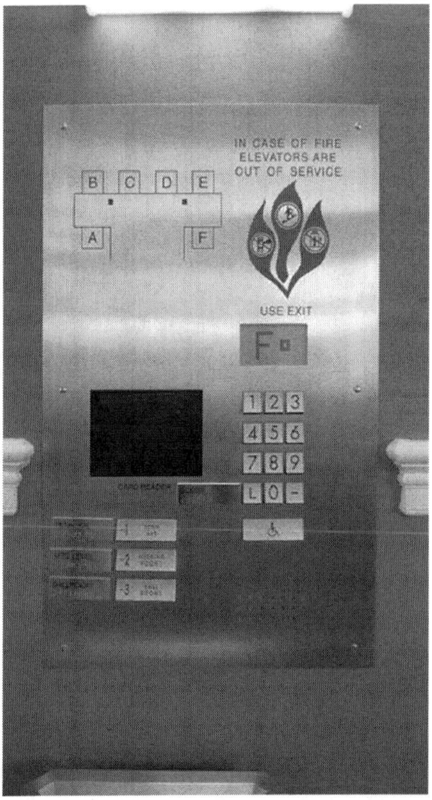


***Figure:*** *A destination dispatch control station, outside of the car, on which the user presses a button to indicate the desired destination floor, and the panel indicates which car will be dispatched*

## Special Operating Modes

***Anti-Crime Protection (ACP):*** Anti-Crime Protection will force each car to stop at a pre-defined landing and open its doors. This allows a security guard or a receptionist at the landing to visually inspect the passengers. The car stops at this landing as it passes to serve further demand.

### Up Peak (MIT)

During Up Peak mode (also called Moderate Incoming Traffic), elevator cars in a group are recalled to the lobby to provide expeditious service to passengers arriving at the building, most typically in the morning as people arrive for work or at the conclusion of a lunch-time period. Elevators are dispatched one-by-one when they reach a pre-determined passenger load, or when they have had their doors opened for a certain period of time. The next elevator to be dispatched usually has its hall lantern or a "this car leaving next" sign illuminated to encourage passengers to make maximum use of the available elevator system capacity.

The commencement of Up Peak may be triggered by a time clock, by the departure of a certain number of fully loaded cars leaving the lobby within a given time period, or by a switch manually operated by a building attendant.

### Down Peak

During Down Peak mode, elevator cars in a group are sent away from the lobby towards the highest floor served, after which they commence running down the floors in response to hall calls placed by passengers wishing to leave the building. This allows the elevator system to provide maximum passenger handling capacity for people leaving the building. The commencement of Down Peak may be triggered by a time clock, by the arrival of a certain number of fully loaded cars at the lobby within a given time period, or by a switch manually operated by a building attendant.

### Sabbath Service (SHO)

In areas with large populations of observant Jews or in facilities catering to Jews, one may find a "Sabbath elevator". In this mode, an elevator will stop automatically at every floor, allowing people to step on and off without having to press any buttons. This prevents violation of the Sabbath prohibition against operating electrical devices when Sabbath is in effect for those who observe this ritual.

However, Sabbath mode has the side effect of using considerable amounts of energy, running the elevator car sequentially up and down every floor of a building, repeatedly servicing floors where it is not needed. For a tall building with many floors, the car must move on a frequent enough basis so as to not cause undue delay for potential users that will not touch the controls as it opens the doors on every floor up the building.

### Independent Service (ISC)

Independent service is a special service mode found on most elevators. It is activated by a key switch either inside the elevator itself or on a centralized control panel in the lobby. When an elevator is placed on independent service, it will no longer respond to hall calls. (In a bank of elevators, traffic is rerouted to the other elevators, while in a single elevator, the hall buttons are disabled). The elevator will remain parked on a floor with its doors open until a floor is selected and the door close button is held until the elevator starts to travel. Independent service is useful when transporting large goods or moving groups of people between certain floors.

### Inspection Service (INS)

Inspection service is designed to provide access to the hoistway and car top for inspection and maintenance purposes by qualified elevator mechanics. It is first activated by a key switch on the car operating panel usually labelled 'Inspection', 'Car Top', 'Access Enable' or 'HWENAB'. When this switch is activated the elevator will come to a stop if moving, car calls will be cancelled (and the buttons disabled), and hall calls will be assigned to other elevator cars in the group (or cancelled in a single elevator configuration). The elevator can now only be moved by the corresponding 'Access' key switches, usually located at the highest (to access the top of the car) and lowest (to access the elevator pit) landings. The access key switches will allow the car to move at reduced inspection speed with the hoistway door open. This speed can range from anywhere up to 60% of normal operating speed on most controllers, and is usually defined by local safety codes.

Elevators have a car top inspection station that allows the car to be operated by a mechanic in order to move it through the hoistway. Generally, there are three buttons — UP, RUN, and DOWN. Both the RUN and a direction button must be held to move the car in that direction, and the elevator will stop moving as soon as the buttons are released. Most other elevators have an up/down toggle switch and a RUN button. The inspection panel also has standard power outlets for work lamps and powered tools.

### *Fire Service Mode (EFS)*

Depending on the location of the elevator, fire service code will vary state to state and country to country. Fire service is usually split up into two modes: Phase One and Phase Two. These are separate modes that the elevator can go into.

Phase one mode is activated by a corresponding smoke sensor or heat sensor in the building. Once an alarm has been activated, the elevator will automatically go into phase one. The elevator will wait an amount of time, then proceed to go into nudging mode to tell everyone the elevator is leaving the floor. Once the elevator has left the floor, depending on where the alarm was set off, the elevator will go to the Fire Recall Floor.

However, if the alarm was activated on the fire recall floor the elevator will have an alternate floor to recall to. When the elevator is recalled, it proceeds to the recall floor and stops with its doors open. The elevator will no longer respond to calls or move in any direction. Located on the fire recall floor is a fire service key switch. The fire service key switch has the ability to turn fire service off, turn fire service on or to bypass fire service. The only way to return the elevator to normal service is to switch it to bypass after the alarms have reset.

Phase two mode can only be activated by a key switch located inside the elevator on the centralized control panel. This mode was created for firefighters so that they may rescue people from a burning building. The phase two key switch located on the COP has three positions: off, on, and hold. By turning phase two on, the firefighter enables the car to move. However, like independent service mode, the car will not respond to a car call unless the firefighter manually pushes and holds the door close button.

Once the elevator gets to the desired floor it will not open its doors unless the firefighter holds the door open button. This is in case the floor is burning and the firefighter can feel the heat and knows not to open the door. The firefighter must hold door open until the door is completely opened. If for any reason the firefighter wishes to leave the elevator, they will use the hold position on the key switch to make sure the elevator remains at that floor. If the firefighter wishes to return to the recall floor, they simply turn the key off and close the doors. Fire Service is for emergency use only, although fire service keys can be purchased on eBay, and other websites. Only trained responders should use this feature, and it is by no means a safe way to escape from a burning building.

### Medical Emergency/Code Blue Service (EHS)

Commonly found in hospitals, Code Blue service allows an elevator to be summoned to any floor for use in an emergency situation. Each floor will have a Code Blue recall key switch, and when activated, the elevator system will immediately select the elevator car that can respond the fastest, regardless of direction of travel and passenger load. Passengers inside the elevator will be notified with an alarm and indicator light to exit the elevator when the doors open.

Once the elevator arrives at the floor, it will park with its doors open and the car buttons will be disabled to prevent a passenger from taking control of the elevator. Medical personnel must then activate the Code Blue key switch inside the car, select their floor and close the doors with the door close button. The elevator will then travel non-stop to the selected floor, and will remain in Code Blue service until switched off in the car. Some hospital elevators will feature a 'hold' position on the Code Blue key switch (similar to fire service) which allows the elevator to remain at a floor locked out of service until Code Blue is deactivated.

### Emergency Power Operation (EPR)

Many elevator installations now feature emergency power systems which allow elevator use in blackout situations and prevent people from becoming trapped in elevators.

### Traction Elevators

When power is lost in a traction elevator system, all elevators will initially come to a halt. One by one, each car in the group will return to the lobby floor, open its doors and shut down. People in the remaining elevators may see an indicator light or hear a voice announcement informing them that the elevator will return to the lobby shortly. Once all cars have successfully returned, the system will then automatically select one or more cars to be used for normal operations and these cars will return to service. The car(s) selected to run under emergency power can be manually overridden by a key or strip switch in the lobby. In order to help prevent entrapment, when the system detects that it is running low on power, it will bring the running cars to the lobby or nearest floor, open the doors and shut down.

### *Hydraulic Elevators*

In hydraulic elevator systems, emergency power will lower the elevators to the lowest landing and open the doors to allow passengers to exit. The doors then close after an adjustable time period and the car remains unusable until reset, usually by cycling the elevator main power switch. Typically, due to the high current draw when starting the pump motor, hydraulic elevators are not run using standard emergency power systems. Buildings like hospitals and nursing homes usually size their emergency generators to accommodate this draw. However, the increasing use of current-limiting motor starters, commonly known as "Soft-Start" contactors, avoid much of this problem, and the current draw of the pump motor is less of a limiting concern.

### *Elevator Modernisation*

Most elevators are built to provide about 20 years of service, as long as service intervals specified and periodic maintenance/inspections by the manufacturer are followed. As the elevator ages and equipment become increasingly difficult to find or replace, along with code changes and deteriorating ride performance, a complete overhaul of the elevator may be suggested to the building owners.

A typical modernisation consists of controller equipment, electrical wiring and buttons, position indicators and direction arrows, hoist machines and motors (including door operators), and sometimes door hanger tracks. Rarely are car slings, rails, or other heavy structures changed. The cost of an elevator modernisation can range greatly depending on which type of equipment is to be installed.

Modernisation can greatly improve operational reliability by replacing mechanical relays and contacts with solid-state electronics. Ride quality can be improved by replacing motor-generator-based drive designs with Variable-Voltage, Variable Frequency (V3F) drives, providing near-seamless acceleration and deceleration. Passenger safety is also improved by updating systems and equipment to conform to current codes.

### *Elevator Safety*

***Cable-Borne Elevators:*** Statistically speaking, cable-borne elevators are extremely safe. Their safety record is unsurpassed by any other vehicle system. In 1998, it was estimated that approximately eight millionths of one percent (1 in 12 million) of elevator rides result in an anomaly, and the vast majority of these were minor things such as the doors failing to open. Of the 20 to 30 elevator-related deaths each year, most of them are maintenance-related — for example, technicians leaning too far into the shaft or getting caught between moving parts, and most of the rest are attributed to other kinds of accidents, such as people stepping blindly through doors that open into empty shafts or being strangled by scarves caught in the doors. In fact, prior to the September 11th terrorist attacks, the only known free-fall incident in a modern cable-borne elevator happened in 1945 when a B-25 bomber struck the Empire State Building in fog, severing the cables of an elevator cab, which fell from the 75th floor all the way to the bottom of the building, seriously injuring (though not killing) the sole occupant — the elevator operator. However, there was an incident in 2007 at a Seattle children's hospital, where a ThyssenKrupp ISIS machine-room-less elevator free-fell until the safety brakes were engaged.

This was due to a flaw in the design where the cables were connected at one common point, and the kevlar ropes had a tendency to overheat and cause slipping (or, in this case, a free-fall). While it is possible (though extraordinarily unlikely) for an elevator's cable to snap, all elevators in the modern era have been fitted with several safety devices which prevent the elevator from simply free-falling and crashing. An elevator cab is typically borne by six or eight hoist cables, each of which is capable on its own of supporting the full load of the elevator plus twenty-five percent more weight.

In addition, there is a device which detects whether the elevator is descending faster than its maximum designed speed; if this happens, the device causes copper brake shoes to clamp down along the vertical rails in the shaft, stopping the elevator quickly, but not so abruptly as to cause injury. This device is called the governor, and was invented by Elisha Graves Otis. In addition, a hydraulic buffer is installed at the bottom of the shaft to somewhat cushion any impact. However, In Thailand, in November 2012, a woman was killed in free falling elevator, in what was reported as the "first legally recognised death caused by a falling lift".

***Hydraulic Elevators:*** Past problems with early hydraulic elevators meant those built prior to a code change in 1972 were subject to possible catastrophic failure. The code had previously required only single-bottom hydraulic cylinders. In the event of a cylinder breach, an uncontrolled fall of the elevator might result. Because it is impossible to verify the system completely without a pressurized casing (as described below), it is necessary to remove the piston to inspect it. The cost of removing the piston is such that it makes no economic sense to re-install the old cylinder; therefore it is necessary to replace the cylinder and install a new piston. Another solution to protect against a cylinder blowout is to install a "life jacket." This is a device which, in the event of an excessive downward speed, clamps onto the cylinder and stops the car. A device known as a rupture valve is often attached to the hydraulic inlet/outlet of the piston and can be adjusted for a maximum flow rate. If a pipe or hose were to break (rupture), the flow rate of the rupture valve will surpass a set limit and mechanically stop the outlet flow of hydraulic fluid, thus stopping the piston and the car in the down direction.

In addition to the safety concerns for older hydraulic elevators, there is risk of leaking hydraulic oil into the aquifer and causing potential environmental contamination. This has led to the introduction of PVC liners (casings) around hydraulic cylinders which can be monitored for integrity. In the past decade, recent innovations in inverted hydraulic jacks have eliminated the costly process of drilling the ground to install a borehole jack. This also eliminates the threat of corrosion to the system and increases safety.

### *Mine-Shaft Elevators*

Safety testing of mine shaft elevator rails is routinely undertaken. The method involves destructive testing of a segment of the cable. The ends of the segment are frayed, then set in conical zincmolds. Each end of the segment is then secured in a large, hydraulic stretching machine. The segment is then placed under increasing load to the point of failure. Data about elasticity, load, and other factors is compiled and a report is produced. The report is then analyzed to determine whether or not the entire rail is safe to use.

### *Passenger Service*

A passenger elevator is designed to move people between a building's floors. Passenger elevators capacity is related to the available floor space. Generally passenger elevators are available in capacities from 1,000 to 6,000 pounds (500–2,700 kg) in 500-pound (230 kg) increments. Generally passenger elevators in buildings of eight floors or fewer are hydraulic or electric, which can reach speeds up to 200 feet per minute (1 m/s) hydraulic and up to 500 feet per minute (152 m/min) electric. In buildings up to ten floors, electric and gearless elevators are likely to have speeds up to 500 feet per minute (3 m/s), and above ten floors speeds range 500 to 2,000 feet per minute (3–10 m/s).

Sometimes passenger elevators are used as a city transport along with funiculars. For example, there is a 3-station underground public elevator in Yalta, Ukraine, which takes passengers from the top of a hill above the Black Sea on which hotels are perched, to a tunnel located on the beach below.

At Casco Viejo station in the Bilbao Metro, the elevator that provides access to the station from a hilltop neighbourhood doubles as city transportation: the station's ticket barriers are set up in such a way that passengers can pay to reach the elevator from the entrance in the lower city, or vice versa.

### Types of Passenger Elevators

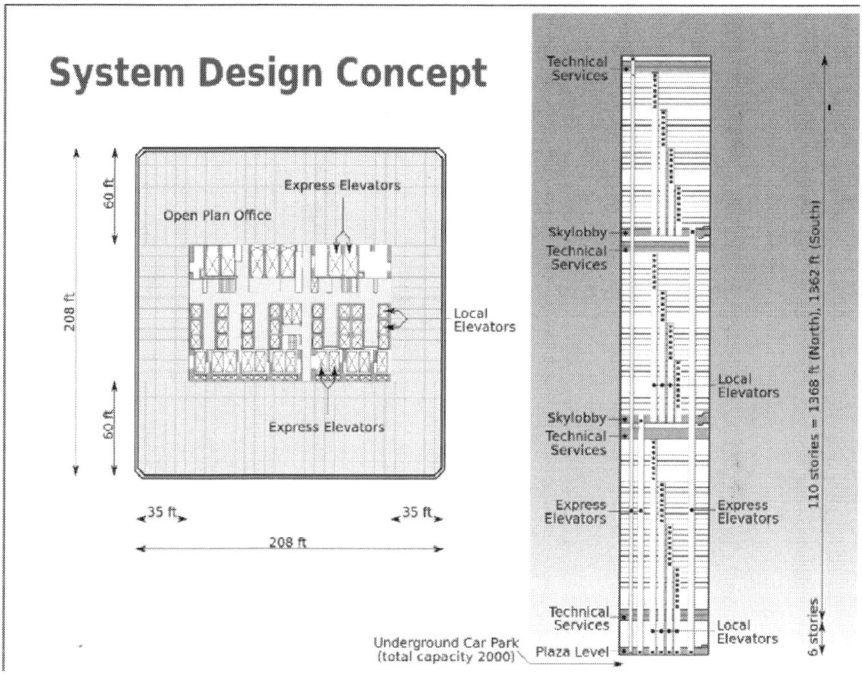


***Figure:** The former World Trade Centre's twin towers used skylobbies, located on the 44th and 78th floors of each tower*

Passenger elevators may be specialised for the service they perform, including: hospital emergency (Code blue), front and rear entrances, a television in high-rise buildings, double decker, and other uses. Cars may be ornate in their interior appearance, may have audio visual advertising, and may be provided with specialised recorded voice announcements. Elevators may also have loudspeakers in them to play calm, easy listening music. Such music is often referred to as elevator music.

An express elevator does not serve all floors. For example, it moves between the ground floor and a skylobby, or it moves from the ground floor or a skylobby to a range of floors, skipping floors in between. These are especially popular in eastern Asia.

### Capacity

Residential elevators may be small enough to only accommodate one person while some are large enough for more than a dozen. Wheelchair, or platform lifts, a specialised type of elevator designed to move a wheelchair 12 feet (3.7 m) or less, can often accommodate just one person in a wheelchair at a time with a load of 750 pounds (340 kg).

### Freight Elevators

A freight elevator, or goods lift, is an elevator designed to carry goods, rather than passengers. Freight elevators are generally required to display a written notice in the car that the use by passengers is prohibited (though not necessarily illegal), though certain freight elevators allow dual use through the use of an inconspicuous riser. In order for an elevator to be legal to carry passengers in some jurisdictions it must have a solid inner door. Freight elevators are typically larger and capable of carrying heavier loads than a passenger elevator, generally from 2,300 to 4,500 kg. Freight elevators may have manually operated doors, and often have rugged interior finishes to prevent damage while loading and unloading. Although hydraulic freight elevators exist, electric elevators are more energy efficient for the work of freight lifting.

### Sidewalk Elevators

A sidewalk elevator is a special type of freight elevator. Sidewalk elevators are used to move materials between a basement and a ground-level area, often the sidewalk just outside the building. They are controlled via an exterior switch and emerge from a metal trap door at ground level. Sidewalk elevator cars feature a uniquely shaped top that allows this door to open and close automatically.

### Stage Lifts

Stage and orchestra lifts are specialised lifts, typically powered by hydraulics, that are used to lift entire sections of a theatre stage. For example, Radio City Music Hall has four such lifts: an "orchestra lift" that covers a large area of the stage, and three smaller lifts near the rear of the stage.

In this case, the orchestra lift is powerful enough to raise an entire orchestra, or an entire cast of performers (including live elephants) up to stage level from below.

***Figure:*** *The pit beneath the orchestra lift at Radio City Music Hall*

### Vehicle Elevators

Vehicular elevators are used within buildings or areas with limited space (in lieu of ramps), typically to move cars into the parking garage or manufacturer's storage. Geared hydraulic chains (not unlike bicycle chains) generate lift for the platform and there are no counterweights. To accommodate building designs and improve accessibility, the platform may rotate so that the driver only has to drive forward. Most vehicle elevators have a weight capacity of 2 tons.

Rare examples of extra-heavy elevators for 20-ton lorries, and even for railcars (like one that was used at Dnipro Station of the Kiev Metro) also occur.

### Boat Elevators

In some smaller canals, boats and small ships can pass between different levels of a canal with a boat lift rather than through a canal lock.

### Elevators for Aircraft

On aircraft carriers, elevators carry aircraft between the flight deck and the hangar deck for operations or repairs. These elevators are designed for much greater capacity than other elevators, up to 200,000 pounds (91,000 kg) of aircraft and equipment. Smaller elevators lift munitions to the flight deck from magazines deep inside the ship.

### Elevators within Aircraft

On some passenger double-deck aircraft such as the Boeing 747 or other widebody aircraft, lifts transport flight attendants and food and beverage trolleys from lower deck galleys to upper passenger carrying decks.

### Limited Use / Limited Application

The limited-use, limited-application (LU/LA) elevator is a special purpose passenger elevator used infrequently, and which is exempt from many commercial regulations and accommodations. For example, a LU/LA is primarily meant to be handicapped accessible, and there might only be room for a single wheelchair and a standing passenger.

### Residential Elevator

A residential elevator is often permitted to be of lower cost and complexity than full commercial elevators. They may have unique design characteristics suited for home furnishings, such as hinged wooden shaft-access doors rather than the typical metal sliding doors of commercial elevators. Construction may be less robust than in commercial designs with shorter maintenance periods, but safety systems such as locks on shaft access doors, fall arrestors, and emergency phones must still be present in the event of malfunction.

The American Society of Mechanical Engineers (ASME) has a specific section of Safety Code (ASME A17.1 Section 5.3) which addresses Residential Elevators. This section allows for different parameters to alleviate design complexity based on the limited use of a residential elevator by a specific user or user group. Section 5.3 of the ASME A17.1 Safety Code is for Private Residence Elevators, which does not include multi-family dwellings.

Some types of residential elevators do not use a traditional elevator shaft, machine room, and elevator hoistway. This allows an elevator to be installed where a traditional elevator may not fit, and simplifies installation. The ASME board first approved machine-room-less systems in a revision of the ASME A17.1 in 2007. Machine-room-less elevators have been available commercially since the mid 1990s, however cost and overall size prevented their adoption to the residential elevator market until around 2010.

***Figure:*** *A residential elevator with integrated hoistway construction and machine-room-less design*

### Dumbwaiter

Dumbwaiters are small freight elevators that are intended to carry food rather than passengers. They often link kitchens with rooms on other floors.

### Paternoster

A special type of elevator is the paternoster, a constantly moving chain of boxes. A similar concept, called the manlift or humanlift, moves only a small platform, which the rider mounts while using a handhold and was once seen in multi-story industrial plants.

### Scissor Lift

The scissor lift is yet another type of lift. As most of these lifts are self-contained, these lifts can be easily moved to where they are needed.

### Rack-and-Pinion Lift

Rack-and-pinion lifts are powered by a motor driving a pinion gear. Because they can be installed on a building or structure's exterior and there is no machine room or hoistway required, they are the most used type of lift for buildings under construction (to move materials and tools up and down).

### Material Handling Belts and Belt Elevators

Material transport elevator generally consist of an inclined plane on which a conveyor belt runs. The conveyor often includes partitions to ensure the material moves forwards. These elevators are often used in industrial and agricultural applications. When such mechanisms (or spiral screws or pneumatic transport) are used to elevate grain for storage in large vertical silos, the entire structure is called a grain elevator. Belt elevators are often used in docks for loading loose materials such as coal, iron ore and grain into the holds of bulk carriers

There have occasionally been lift belts for humans; these typically have steps about every 7 feet (2.1 m) along the length of the belt, which moves vertically, so that the passenger can stand on one step and hold on to the one above. These belts are sometimes used, for example, to carry the employees of parking garages, but are considered too dangerous for public use.

### Elevator Convenience Features

Elevators may feature talking devices as an accessibility aid for the blind. In addition to floor arrival notifications, the computer announces the direction of travel, and notifies the passengers before the doors are to close.

In addition to the call buttons, elevators usually have floor indicators (often illuminated by LED) and direction lanterns. The former are almost universal in cab interiors with more than two stops and may be found outside the elevators as well on one or more of the floors. Floor indicators can consist of a dial with a rotating needle, but the most common types are those with successively illuminated floor indications or LCDs. Likewise, a change of floors or an arrival at a floor is indicated by a sound, depending on the elevator.

***Figure:*** *LCD elevator floor indicator*

***Figure:*** *A typical Elevator Indicator located in the Waldorf Astorria Hotel in New York City NY. This elevator was made by Otis*

Direction lanterns are also found both inside and outside elevator cars, but they should always be visible from outside because their primary purpose is to help people decide whether or not to get on the elevator. If somebody waiting for the elevator wants to go up, but a car comes first that indicates that it is going down, then the person may decide not to get on the elevator.

If the person waits, then one will still stop going up. Direction indicators are sometimes etched with arrows or shaped like arrows and/or use the convention that one that lights up red means "down" and green means "up". Since the colour convention is often undermined or overridden by systems that do not invoke it, it is usually used only in conjunction with other differentiating factors. An example of a place whose elevators use only the colour convention to differentiate between directions is the Museum of Contemporary Art in Chicago, where a single circle can be made to light up green for "up" and red for "down". Sometimes directions must be inferred by the position of the indicators relative to one another.

In addition to lanterns, most elevators have a chime to indicate if the elevator is going up or down either before or after the doors open, usually in conjunction with the lanterns lighting up. Universally, one chime is for up, two is for down, and none indicates an elevator that is 'free'. Observatory service elevators often convey other facts of interest, including elevator speed, stopwatch, and current position (altitude), as with the case for Taipei 101's service elevators.

### Elevator Air Conditioning

### Concept

Elevator air conditioning is fast becoming a popular concept around the world. The primary reason for installing an elevator air conditioner is the comfort that it provides while travelling in the elevator. It stabilizes the condition of the air inside the lift car. Some elevator air conditioners can be used in countries with cold climates if a thermostat is used to reverse the refrigeration cycle to warm the lift car.

### Health

One of the benefits of installing an elevator air conditioner is the clean air it provides. Air has typically been drawn from the elevator shaft or hoistway into the car using a motorized fan, which may not be especially clean air. With an elevator air conditioner, the air is much cleaner because it is recirculated within the car itself and is usually filtered to remove contaminants. A poorly maintained air-conditioning system may promote the growth and spread of microorganisms, but as long as the air conditioner is kept clean, these health hazards can be avoided.

### Drawbacks

Heat generated from the cooling process is dissipated into the hoistway. The elevator cab (or car) is ordinarily not air-tight, and some of this heat may reenter the car and reduce the overall cooling effect.

### Energy

The air from the lobby constantly leaks into the elevator shaft due to elevator movements as well as elevator shaft ventilation requirements. Using this conditioned air in the elevator does not increase energy costs. However, by using an independent elevator air conditioner to achieve better temperature control inside the car, more energy will be used.

### Condensation

Air conditioning poses a problem to elevators because of the condensation that occurs. The condensed water produced has to be disposed of; otherwise, it would create flooding in the elevator car and hoistway.

### Methods of Removing Condensed Water

There are at least four ways to remove condensed water from the air conditioner. However, each solution has its pros and cons.

### Atomizing

Atomizing, also known as misting the condensed water, is another way to dispose of the condensed water. Spraying ultra-fine water droplets onto the hot coils of the air conditioner ensures that the condensed water evaporates quickly.

Though this is one of the best methods to dispose of the condensed water, it is also one of the costliest because the nozzle that atomizes the water easily gets choked. The majority of the cost goes to maintaining the entire atomizing system.

### Boiling

Disposing of condensed water works by firstly collecting the condensed water and then heating it to above boiling point. The condensed water is eventually evaporated, thereby disposing of it. Consumers are reluctant to employ this system because of the high rate of energy used just to dispose of this water.

### Cascading

The cascading method works by flowing the condensed water directly onto the hot coils of the air conditioner. This eventually evaporates the condensed water. The downside of this technology is that the coils have to be at extremely high temperature for the condensed water to be evaporated. There is a chance that the water might not evaporate entirely and that would cause water to overflow onto the exterior of the car.

### Drainage System

Drainage system works by creating a sump to collect the condensed water and using a pump to dispose it through a drainage system.

It is an efficient method, but it comes at a heavy price because the cost of building the sump. Moreover, maintaining the pump to make sure it operates is very expensive. Furthermore, the pipes used for drainage would look ugly on the exterior. This system also cannot be implemented on a built project.

### Standards

The mechanical and electrical design of elevators is dictated according to various standards (aka elevator codes), which may be international, national, state, regional or city based. Whereas once many standards were prescriptive, specifying exact criteria which must be complied with, there has recently been a shift towards more performance-based standards where the onus falls on the designer to ensure that the elevator meets or exceeds the standard.

Some of the national elevator standards include:

- Australia – AS1735
- Canada – CAN/CSA B44
- Europe – EN 81 series (EN 81-1, EN 81-2, EN 81-28, EN 81-70, EN 12015, EN 12016, EN 13015, etc.)
- USA – ASME A17

Because an elevator is part of a building, it must also comply with standards relating to earthquake resilience, fire standards, electrical wiring rules and so forth.

The American National Elevator Standards Group (ANESG) sets an elevator weight standard to be 2,200 pounds (1,000 kg).

Additional requirements relating to access by disabled persons, may be mandated by laws or regulations such as the Americans with Disabilities Act.

### U.S. and Canadian Elevator Standard Specifics

In most US and Canadian jurisdictions, passenger elevators are required to conform to the American Society of Mechanical Engineers' Standard A17.1, Safety Code for Elevators and Escalators. As of 2006, all states except Kansas, Mississippi, North Dakota, and South Dakota have adopted some version of ASME codes, though not necessarily the most recent. In Canada the document is the CAN/CSA B44 Safety Standard, which was harmonized with the US version in the 2000 edition. In addition, passenger elevators may be required to conform to the requirements of A17.3 for existing elevators where referenced by the local jurisdiction. Passenger elevators are tested using the ASME A17.2

Standard. The frequency of these tests is mandated by the local jurisdiction, which may be a town, city, state or provincial standard.

Passenger elevators must also conform to many ancillary building codes including the Local or State building code, National Fire Protection Association standards for Electrical, Fire Sprinklers and Fire Alarms, Plumbing codes, and HVAC codes. Also, passenger elevators are required to conform to the Americans with Disabilities Act and other State and Federal civil rights legislation regarding accessibility.

Residential elevators are required to conform to ASME A17.1. Platform and Wheelchair lifts are required to comply with ASME A18.1 in most US jurisdictions.

Most elevators have a location in which the permit for the building owner to operate the elevator is displayed. While some jurisdictions require the permit to be displayed in the elevator cab, other jurisdictions allow for the operating permit to be kept on file elsewhere – such as the maintenance office – and to be made available for inspection on demand. In such cases instead of the permit being displayed in the elevator cab, often a notice is posted in its place informing riders of where the actual permits are kept.

## Unique Elevator Installations

***Table:*** *World statistics*

| *Country* | *Number of Elevators Installed* |
|---|---|
| Italy | 900,000 |
| United States | 700,000 |
| People's Republic of China | 610,000 |
| Russia | 520,000 |

As of January 2008, Italy is the nation with the most elevators installed in the world, with 850,000 elevators installed that run more than one hundred million lifts every day, followed by United States with 700,000 elevators installed and People's Republic of China with 610,000 elevators installed since 1949. In Brazil, it is estimated that there are approximately 300,000 elevators currently in operation. The world's largest market for elevators is Italy, with more than 1,629 million euros of sales and 1,224 million euros of internal market.

## Eiffel Tower

The Eiffel Tower has Otis double-deck elevators built into the legs of the tower, serving the ground level to the first and second levels. Even though the shaft runs diagonally upwards with the contour of the tower, both the upper and lower cars remain horizontally level. The offset distance of the two cars changes throughout the journey.

There are four elevator cars of the traditional design that run from the second level to the third level. The cars are connected to their opposite pairs (opposite in the elevator landing/hall) and use each other as the counterweight. As one car ascends from level 2, the other descends from level 3. The operations of these elevators are synchronized by a light signal in the car.

### *Taipei 101*

Double deck elevators are used in the Taipei 101 office tower. Tenants of even-numbered floors first take an escalator (or an elevator from the parking garage) to the 2nd level, where they will enter the upper deck and arrive at their floors. The lower deck is turned off during low-volume hours, and the upper deck can act as a single-level elevator stopping at all adjacent floors. For example, the 85th floor restaurants can be accessed from the 60th floor sky-lobby. Restaurant customers must clear their reservations at the reception counter on the 2nd floor. A bank of express elevators stop only on the sky lobby levels (36 and 60, upper-deck car), where tenants can transfer to "local" elevators.

The high-speed observation deck elevators accelerate to a world-record certified speed of 1,010 metre per minute (61 km/h) in 16 seconds, and then it slows down for arrival with subtle air pressure sensations. The door opens after 37 seconds from the 5th floor. Special features include aerodynamic car and counterweights, and cabin pressure control to help passengers adapt smoothly to pressure changes. The downwards journey is completed at a reduced speed of 600 metres per minute, with the doors opening at the 52nd second.

### *The Gateway Arch*

The Gateway Arch in St. Louis, Missouri has a unique elevator system which carries passengers from the visitors' centre underneath the Arch to the observation deck at the top of the structure. Called a *tram* or *tramway*, people enter this unique tramway much as one would enter an ordinary elevator, through double doors. Passing through the doors the passengers in small groups enter a horizontal cylindrical compartment containing seats on each side and a flat floor. A number of these compartments are linked to form a train. These compartments each individually retain an appropriate level orientation by tilting while the entire train follows curved tracks up one leg of the arch.

There are two tramways within the Arch, one at the north end, and the other at the south end. The entry doors have windows, so people travelling within the Arch are able to see the interior structure of the Arch during the ride to and from the observation deck. At the beginning of the trip the cars hang from the drive cables, but as the angle of the shaft changes, they end up beside and then on top of the cables.

### *New City Hall, Hanover, Germany*

The elevator in the New City Hall in Hanover, Germany is a technical rarity, and unique in Europe, as the elevator starts straight up but then changes its angle by 15 degrees to follow the contour of the dome of the hall. The cabin therefore tilts 15 degrees during the ride. The elevator travels a height of 43 metres. The new city hall was built in 1913. The elevator was destroyed in 1943 and rebuilt in 1954.

### *Luxor Incline Elevator*

In Las Vegas, Nevada, at the Luxor Hotel, the elevators are Inclined elevators. The shape of this casino is a pyramid. Therefore, the elevator travels up the side of the pyramid at a 39-degree angle. Other locations with inclined elevators include the Cityplace Station in Dallas, Texas, the Huntington Metro Station in Huntington, Virginia, and the San Diego Convention Centre in San Diego, California.

### *Twilight Zone Tower of Terror*

The Twilight Zone Tower of Terror is the common name for a series of elevator attractions at the Disney's Hollywood Studios park in Orlando, the Disney California Adventure Parkpark in Anaheim, the Walt Disney Studios Park in Paris and the Tokyo DisneySea park in Tokyo. The central element of this attraction is a simulated free-fall achieved through the use of a high-speed elevator system. For safety reasons, passengers are seated and secured in their seats rather than standing. Unlike most traction elevators, the elevator car and counterweight are joined using a rail system in a continuous loop running through both the top and the bottom of the drop shaft. This allows the drive motor to pull down on the elevator car from underneath, resulting in downward acceleration greater than that of normal gravity. The high-speed drive motor is used to rapidly lift the elevator as well.

The passenger cabs are mechanically separated from the lift mechanism, thus allowing the elevator shafts to be used continuously while passengers board and embark from the cabs. Multiple elevator shafts are used to further improve passenger throughput. The doorways of the top few "floors" of the attraction are open to the outdoor environment, thus allowing passengers to look out from the top of the structure.

### "Top of the Rock" Elevators

Guests ascending to the 67th, 69th, and 70th level observation decks (dubbed "Top of the Rock") atop the GE Building at Rockefeller Centre in New York City ride a high-speed glass-top elevator. When entering the cab, it appears to be any normal elevator ride. However, once the cab begins moving, the interior lights turn off and a special blue light above the cab turns on. This lights the entire shaft, so riders can see the moving cab through its glass ceiling as it rises and lowers through the shaft. Music plays and various animations are also displayed on the ceiling. The entire ride takes about 60 seconds.

### The Haunted Mansion

Part of the Haunted Mansion attraction at Disneyland in Anaheim, California and Disneyland in Paris, France, takes place on an elevator. The "stretching room" on the ride is actually an elevator that travels downwards, giving access to a short underground tunnel which leads to the rest of the attraction. The elevator has no ceiling and its shaft is decorated to look like walls of a mansion. Because there is no roof, passengers are able to see the walls of the shaft by looking up, which gives the illusion of the room stretching.

## Fireplace

A fireplace is an architectural structure designed to contain a fire. Fireplaces are used at the present time mostly for the relaxing ambiance they create. Historically they were used for the practical purposes of heating, cooking, and heating water for laundry and domestic uses. A fire is contained in a firebox or firepit; a chimney or other flue allows exhaust to escape. A fireplace may have: a foundation; a hearth; a firebox; a mantelpiece; a chimney crane, used in kitchen and laundry fireplaces; a grate; a lintel; a lintel bar; overmantel; a damper; a smoke chamber; a throat; a flue. On the exterior there is often a corbeled brick crown; the projecting courses of brick act as a drip course to keep rainwater from running down the exterior walls. A cap, hood, or shroud serves to keep rainwater out of the exterior of the chimney; rain in the chimney is a much greater problem in chimneys lined with impervious flue tiles or metal liners than with the traditional masonry chimney, which soaks up all but the most violent rain. Some chimneys have a spark arrestor incorporated into the crown or cap. Fireplaces have variable heat efficiency. Organisations like the Environmental Protection Agency and the Washington Department of Ecology warn that, according to various studies, fireplaces can pose a significant health risk. The EPA writes "Smoke may smell good, but it's not good for you."

### *Types of Fireplaces*

- Masonry fireplaces are made from brick or stone.
- Manufactured fireplaces are made with sheet metal fire boxes.

Masonry and prefabricated fireplaces can be fuelled with wood, natural gas, biomass and propane fuel sources. Ventless Fireplaces (duct free/room-venting fireplaces) are fuelled by either gel, liquid propane, bottled gas or natural gas. In the US, some states and local counties have laws restricting these types of fireplaces. They must be sized appropriately to the area to be heated. There are also air quality control issues due to the amount of moisture they release into the room air, and oxygen sensor and carbon monoxide sensors are safety essentials.

Chimney and flue types:

- Masonry (brick or stone fireplaces and chimneys) with or without tile-lined flue.
- Reinforced concrete chimneys. Fundamental design flaws bankrupted the US manufacturers and made the design obsolete. These chimneys often show vertical cracks on the exterior.
- Metal-lined flue: Double or triple walled metal pipe running up inside a new or existing wood-framed or masonry chase.

Newly constructed flues may feature a chase cover, a cap, and a spark arrestor at the top to keep small animals out and to prevent sparks from being broadcast into the atmosphere.

### *Accessories*

A wide range of accessories are used with fireplaces, which vary between countries and regions, and historical periods. For the interior, common in recent Western cultures are grates, fireguards, log boxes, andirons, pellet baskets, and fire dogs, all of which cradle fuel and accelerate burning. Heavy metal firebacks are sometimes used to capture and re-radiate heat, to protect the back of the fireplace, and as decoration. Fenders are low metal frames set in front of the fireplace to contain embers, soot and ash. For fireplace tending, tools include pokers, bellows, tongs, shovels, brushes and tool stands.

### History

Ancient fire pits were sometimes built in the ground, within caves, or in the centre of a hut or dwelling. Evidence of prehistoric, man-made fires exists on all five continents. The disadvantage of early indoor fire pits was that they produced toxic and/or irritating smoke inside the dwelling.

Fire pits developed into raised hearths in buildings, but venting smoke depended on open windows or holes in roofs. The medieval great hall typically had a centrally located hearth, where an open fire burned with the smoke rising to the vent in the roof. Louvers were developed during the Middle Ages to allow the roof vents to be covered so rain and snow would not enter.

Also during the Middle Ages, smoke canopies were invented to prevent smoke from spreading through a room and vent it out through a wall or roof. These could be placed against stone walls, instead of taking up the middle of the room, and this allowed smaller rooms to be heated.

Chimneys were invented in northern Europe in the 11th or 12th centuries and largely fixed the problem of fumes, more reliably venting smoke outside. They made it possible to give the fireplace a draft, and also made it possible to put fireplaces in multiple rooms in buildings conveniently. They did not come into general use immediately, however, as they were expensive to build and maintain.

In 1678 Prince Rupert, nephew of Charles I, raised the grate of the fireplace, improving the airflow and venting system. The 18th century saw two important developments in the history of fireplaces. Benjamin Franklin developed a convection chamber for the fireplace that greatly improved the efficiency of fireplaces and wood stoves. He also improved the airflow by pulling air from a basement and venting out a longer area at the top. In the later 18th century, Count Rumford designed a fireplace with a tall, shallow firebox that was better at drawing the smoke up and out of the building. The shallow design also improved greatly the amount of radiant heat projected into the room. Rumford's design is the foundation for modern fireplaces.

### Evolution of Fireplace Design

Over time the design of fireplaces has changed from one of necessity to one of visual interest. Early ones were more fire pits than modern fireplaces. They were used for warmth on cold days and nights, as well as for cooking. They also served as a gathering place within the home. These fire pits were usually centred within a room, allowing more people to gather around it.

Many flaws were found in early fireplace designs. Along with the industrial revolution came large scale housing developments, obviating a standardization of fireplaces. The most renowned fireplace designers of this time were the Adam Brothers. They perfected a style of fireplace design that was used for generations. It was smaller, more brightly lit, with an emphasis on the quality of the materials used in their construction, instead of their size. By the 1800s most new fireplaces were made up of two parts, the surround and the insert. The surround consisted of the mantlepiece and sides supports, usually in wood, marble or granite. The insert was where the fire burned, and was constructed of cast iron often backed with decorative tiles. As well as providing heat, the fireplaces of the Victorian era were thought to add a cozy ambiance to homes.

### Heating Efficiency

Some fireplace units incorporate a blower which transfers more of the fireplace's heat to the air via convection, resulting in a more evenly heated space and a lower heating load. Fireplace efficiency can also be increased with the use of a fireback, a piece of metal that sits behind the fire and reflects heat back into the room. Firebacks are traditionally made from cast iron, but are also made from stainless steel.

Most older fireplaces have a relatively low efficiency rating. Standard, modern, wood-burning masonry fireplaces have an efficiency rating of as high as 15%. To improve efficiency, fireplaces can be modified by inserting special heavy fireboxes designed to burn much cleaner and can reach efficiencies as high as 80%. These modified fireplaces are often equipped with a large fire window, enabling an efficient heating process in two phases. During the first phase the initial heat is provided through a large glass window while the fire is burning.

During this time the structure, built of refractory bricks, absorbs the heat. This heat is then evenly radiated for many hours during the second phase. Masonry fireplaces without a glass fire window only provide heat radiated from its surface. Depending on outside temperatures 1 to 2 daily firings are sufficient to ensure a constant room temperature.

### *Health Effects*

In a literature review published in the journal of Toxicology and Environmental Health, J.T. Zelikoff concludes that there are a wide variety of health risks posed by residential wood combustion. She writes:

"With regard to adults, studies show that prolonged inhalation of wood smoke contributed to chronic bronchitis, chronic interstitial lung disease, pulmonary arterial hypertension and corpulmonale, and altered pulmonary immune defence mechanisms. While adverse effects on adults are notable, children appear to be at greatest risk. Many studies that focused specifically on RWC [*Residential Wood Combustion*] have concluded that young children living in homes heated by a wood-burning stove had a greater occurrence of moderate and severe chronic respiratory symptoms than children of the same age and sex who did not live in homes heated with a wood burning stove.

Exposure of preschool children living in homes heated with wood burning stoves or in houses with open fireplaces yielded these effects: decreased pulmonary lung function in young asthmatics; increased incidence of acute bronchitis and severity/frequency of wheezing and coughing; and increased incidence, duration, and possibly severity of acute respiratory infections. Residential wood combustion emissions also contain sulfur oxides, nitrogen oxides, carbon monoxide and potentially carcinogenic compounds including polycyclic aromatic hydrocarbons, benzene, formaldehyde and dioxins. Some of these pollutants are known to cause cancer but their effects on human health via exposure to wood smoke have not been extensively studied."

The Washington Department of Ecology also published a booklet explaining why wood smoke can be dangerous. It explains that human lung and respiratory systems cannot filter the particles emitted by wood combustion, which penetrate deeply into the lungs. For months, cancer causing chemicals can continue to cause changes and structural damage within the respiratory system. Young children, seniors, pregnant women, smokers and individuals with respiratory disorders are most vulnerable. Wood smoke can cause disease and even death in children because it is associated with lower respiratory tract infections.

## Floor

A floor is the walking surface of a room or vehicle. Floors vary from simple dirt in a cave to many-layered surfaces using modern technology. Floors may be stone, wood, bamboo, metal, or any other material that can hold a person's weight. The levels of a building are often referred to as floors although a more proper term is storey.

Floors typically consist of a subfloor for support and a floor covering used to give a good walking surface. In modern buildings the subfloor often has electrical wiring, plumbing, and other services built in. Because floors meet many needs, some essential to safety, floors are built to strict building codes in the first world.

### Special Floor Structures

Where a special floor structure like a floating floor is laid upon another floor then both may be referred to as subfloors.

Special floor structures are used for a number of purposes:

- Balcony, a platform projecting from a wall
- Floating floor, normally for noise or vibration reduction
- Glass floor, as in glass bottomed elevators
- Nightingale floor makes a noise when an intruder walks on it
- Raised floor, utilities underneath can be accessed easily
- Sprung floor, improves the performance and safety of athletes and dancers

### Subfloor Construction

Floors may be built on beams or joists or use structures like prefabricated hollow core slabs. The subfloor builds on those and attaches by various means particular to the support structure but the support and subfloor together always provides the strength of a floor one can sense underfoot. Nowadays, subfloors are generally made from at least two layers of moisture resistant ('AC' grade, one side finished and sanded flat) plywood or composite sheeting, jointly also termed *Underlayments* on floor joists of 2x8, 2x10, or 2x12's (dimensional lumber) spaced generally on 16-inch (40.6 cm) centers, in the United States and Canada.

Some flooring components used solely on concrete slabs consist of a dimpled rubberized or plastic layer much like bubble wrap that provide little tiny pillars for the one-half-inch (12.7 mm) sheet material above. These are manufactured in 2 ft × 2 ft (61 cm × 61 cm) squares and the edges fit together like a mortise and tenon joint. Like a floor on joists not on concrete, a second sheeting underlayment layer is added with staggered joints to disperse forces that would open a joint under the stress of live loads like a person walking.

Three layers are common only in high end highest quality construction. The two layers in high quality construction will both be thick $^{3}/_{4}$ inches (19.1 mm) sheets (as will the third when present), but the two layers may achieve a combined thickness of only half-that in cheaper construction — $^{1}/_{2}$ in (12.7 mm) panel overlaid by $^{1}/_{4}$ in (6.4 mm) plywood subflooring. At the highest end, or in select rooms of the building there might well be three sheeting layers, and such stiff subflooring is necessary to prevent the cracking of large floor tiles of 9–10 inches (22.9–25.4 cm) or more on a side, and the structure under such a floor will frequently also have extra 'bracing' and 'blocking' joist-to-joist intended spread the weight to have as little sagging on any joist as possible when there is a live load on the floor above.

In Europe and North America only a few rare floors will be seen to have no separate floor covering on top, and those are normally because of a temporary condition pending sales or occupancy; in semi-custom new construction and some rental markets, such floors are provided for the new home buyer (renter) to select their own preferred floor coverings usually a wall to wall carpet, or one piece vinyl floor covering. Wood clad ('Hardwood') and tile covered finished floors generally will require a stiffer higher quality subfloor, especially for the later class. Since the wall base and flooring interact forming a joint, such later added semi-custom floors will generally not be hardwood for that joint construction would be in the wrong order unless the wall base trim was also delayed pending the choosing.

The subfloor may also provide underfloor heating and if floor radiant heating is not used, will certainly suffer puncture openings to be put through for forced air ducts for both heating and air conditioning, or pipe holes for *forced hot water* or *steam heating* transport piping conveying the heat from furnace to the to local room's heat exchangers (radiators).

Some sub-floors are inset below the top surface level of surrounding flooring's joists and such subfloors and a normal height joist are joined to make a plywood box both molding and containing at least two inches (5 cm) of concrete (A 'Mud Floor' in builders parlance). Alternatively, only a slightly inset floor topped by a fibrous mesh and concrete building composite floor cladding is used for smaller high quality tile floors—these 'concrete' subfloors have a good thermal match with ceramic tiles and so are popular with builders constructing kitchen, laundry and especially both common and high end bathrooms and any other room where large expanses of well supported ceramic tile will be used as a finished floor. Floors using small (4.5 in or 11.4 cm and smaller) ceramic tiles generally use only an additional $^1/_4$-inch (6.4 mm) layer of plywood (if that) and substitute adhesive and substrate materials making do with both a flexible joints and semi-flexible mounting compounds and so are designed to withstand the greater flexing which large tiles cannot tolerate without breaking.

### Ground Floor Construction

A ground-level floor can be an earthen floor made of soil, or be solid ground floors made of concrete slab.

Ground level slab floors are uncommon in northern latitudes where freezing provides significant structural problems, except in heated interior spaces such as basements or for outdoor unheated structures such as a gazebo or shed where unitary temperatures are not creating pockets of troublesome meltwaters. Ground-level slab floors are prepared for pouring by grading the site, which usually also involves removing topsoil and other organic materials well away from the slab site. Once the site has reached a suitable firm inorganic base material that is graded further so that it is flat and level, and then topped by spreading a layer-cake of force dispersing sand and gravel. Deeper channels may be dug, especially the slab ends and across the slab width at regular intervals in which a continuous run of rebar is bent and wired to sit at two heights within forming a sub-slab 'concrete girder'. Above the targeted bottom height (coplanar with the compacted sand and gravel topping) a separate grid of rebar or welded wire mesh is usually added to reinforce the concrete, and will be tied to the under slab 'girder' rebar at intervals. The under slab cast girders are used especially if it the slab be used structurally, i.e. to support part of the building.

### Upper Floor Construction

Floors in woodframe homes are usually constructed with joists centred no more than 16 inches (41 centimetres) apart, according to most building codes. Heavy floors, such as those made of stone, require more closely spaced joists. If the span between load-bearing walls is too long for joists to safely support, then a heavy crossbeam (thick or laminated wood, or a metal I-beam or H-beam) may be used. A "subfloor" of plywood or waferboard is then laid over the joists.

### Utilities

In modern buildings, there are numerous services provided via ducts or wires underneath the floor or above the ceiling. The floor of one level typically also holds the ceiling of the level below (if any).

Services provided by subfloors include:

- Air conditioning
- Communication fibres
- Electrical wiring
- Fire protection
- Thermal insulation
- Plumbing
- Sewerage
- Soundproofing
- Underfloor heating

In floors supported by joists, utilities are run through the floor by drilling small holes through the joists to serve as conduits. Where the floor is over the basement or crawlspace, utilities may instead be run under the joists, making the installation less expensive. Also, ducts for air conditioning (central heating and cooling) are large and cannot cross through joists or beams; thus, ducts are typically at or near the plenum, or come directly from underneath (or from an attic).

Pipes for plumbing, sewerage, underfloor heating, and other utilities may be laid directly in slab floors, typically via cellular floor raceways. However, later maintenance of these systems can be expensive, requiring the opening of concrete or other fixed structures. Electrically heated floors are available, and both kinds of systems can also be used in wood floors as well.

### *Issues with Floors*

Wood floors, particularly older ones, will tend to 'squeak' in certain places. This is caused by the wood rubbing against other wood, usually at a joint of the subfloor. Firmly securing the pieces to each other with screws or nails may reduce this problem. Floor vibration is a problem with floors. Wood floors tend to pass sound, particularly heavy footsteps and low bass frequencies. Floating floors can reduce this problem. Concrete floors are usually so massive they do not have this problem, but they are also much more expensive to construct and must meet more stringent building requirements due to their weight.

The flooring may need protection sometimes e.g. a gym floor alternately used for seating in a graduation ceremony. A gym floor cover can be used to reduce the need to satisfy incompatible requirements.

# 4

# Flooring

Flooring is the general term for a permanent covering of a floor, or for the work of installing such a floor covering. Floor covering is a term to generically describe any finish material applied over a floor structure to provide a walking surface. Both terms are used interchangeably but floor covering refers more to loose-laid materials.

Materials almost always classified as floor covering include carpet, area rugs, and resilient flooring such as linoleum or vinyl flooring. Materials commonly called flooring include wood flooring, ceramic tile, stone, terrazzo, and various seamless chemical floor coatings.

## Subfloor

The floor under the flooring is called the subfloor. This provides the support for the flooring. Special purpose subfloors like floating floors, raised floors or sprung floors may be laid upon another underlying subfloor which provides the structural strength.

### *Flooring Materials*

The choice of material for floor covering is affected by factors such as cost, endurance, noise insulation, comfort and cleaning effort. Some types of flooring must not be installed below grade (lower than ground level), and laminate or hardwood should be avoided where there may be moisture or condensation.

The sub-floor may be finished in a way that makes it usable without any extra work, see:

- Earthen floor adobe or clay floors
- Solid ground floor cement screed or granolithic

***Figure:*** *Example of stone flooring : white marble slabs covering the floor of the courtyard of the Mosque of Uqba also known as the Great Mosque of Kairouan, in Tunisia.*

### Soft Coverings

Carpet is a floor covering woven or felted from natural or man-made fibres. Fitted carpet is attached to the floor structure, extends wall-to-wall, and cannot be moved from place to place. An underlay will extend carpet life and improve comfort.

Rugs are also woven or felted from fibres, but are smaller than the room in which they are located, have a finished edge, and usually lie over another finished floor such as wood flooring. Rugs may either be temporarily attached to the flooring below by adhesive tape or other methods to prevent creep, or may be loose-laid.

### Wood Flooring

Many different species of wood are fabricated into wood flooring in two primary forms: plank and parquet. Hardwoods are typically much more durable than softwoods. Reclaimed lumber has a unique appearance and is used in green (environmentally responsible) building.

Laminate is a floor covering that appears similar to hardwood but is made with a plywood or medium density fibreboard ("MDF") core with a plastic laminate top layer. HDF laminate consists of high density fibreboard topped by one or more layers of decorative paper and a transparent protective layer. Laminate may be more durable than hardwood, but cannot be refinished like hardwood. Laminate flooring is available in many different patterns which can resemble different woods or even ceramic tile. It usually locks or taps together.

Bamboo flooring is a floor manufactured from the bamboo plant and is a type of hardwood flooring, though technically not a wood. Bamboo is known to be durable and environmentally friendly. It is available in many different patterns, colours, and textures.

Cork Flooring is a flooring material manufactured from the by-product of the cork oak tree. Cork floors are considered to be eco-friendly since the cork oak tree bark is stripped every nine to ten years and doesn't damage the tree. Cork flooring comes in both tiles and planks, and can have glue or glues-less installation.

### Resilient Flooring

Resilient flooring is made of material that has some elasticity. It includes many different manufactured products which including linoleum, sheet vinyl, vinyl composition tile (VCT), cork (sheet or tile), rubber, and others. Performance surfaces used for dance or athletics are made of either wood or resilient flooring.

### Hard Flooring

Ceramic tile includes a wide variety of clay products fired into thin units which are set in beds of mortar or mastic with the joints between tiles grouted. Varieties include quarry tile, porcelain tile, terracotta tile, and others. Many different natural stones are cut into a variety of sizes, shapes, and thicknesses for use as flooring. Stone flooring is usually set in mortar and grouted similar to ceramic tile. Slate and marble are popular types of stone flooring.

New technologies are emerging since 2004 to produce hard floorings having the ability to light up when needed. These security glazing materials contain transparent LED embedded films laminated between glass.

Terrazzo consists of marble or other stone aggregate set in mortar and ground and polished to a smooth surface.

A mosaic consists of many small pieces of stone arranged to form a design or picture and are common in cultural and religious centres.

Concrete/cement finished floor also in popular demand nowadays, for its ability to be treated for different feel and the durability which can last for lifetime.

### Seamless Chemical Flooring

Many different seamless flooring materials are available. These are usually latex, polyester, urethane or epoxy compounds which are applied in liquid form to provide a completely seamless floor covering. These are usually found in wet areas such as laboratories or food processing plants. These may have granular or rubberized particles added to give better traction.

### Sustainable Flooring

Sustainable flooring is produced from sustainable materials (and by a sustainable process) that reduces demands on ecosystems during its life-cycle. Some think that sustainable flooring creates safer and healthier buildings and guarantees a future for traditional producers of renewable resources that many communities depend on.

### Flooring Tools

Special tools used for flooring include:

- Flooring clamp, a clamp for tongue-and-groove floors while nailing
- Knee kicker, used to position carpets precisely and stretch small areas, like steps
- Concrete moisture metre used to check a concrete floor before laying flooring on top

### Floor Features

There are a number of special features that may be used to ornament a floor or perform a useful service:

- Floor medallions decorative centrepieces of a floor design
- Doormats to help keep a floor clean
- Gratings used to drain water or to rub dirt off shoes
- Tactile or rumble strips to warn of for instance a wheelchair ramp, these would normally also be distinctively coloured or patterned.
- Light strips to show an escape route out, especially on airplanes.
- Moldings or baseboards to decorate the sides of a floor or to cover the edge of a floating floor.
- Anti-slip mats: The addition of either granular or rubberized particles that will allow wheels, shoes, or feet better traction.

### Issues with Hardwood Flooring

Hardwood flooring needs to be installed on a sound, flat subfloor (within 1/8" over 6'), and in a controlled environment. Repeated and extreme changes in temperature and/or humidity causing the wood to expand and contract will stress the product and lead to several issues such as squeaking, warping, cupping, crowning, gapping, and delamination. A squeaky floor is generally a symptom of the floor not being securely attached to a wood subfloor. Or, that the wood subfloor is not properly secured to the joists below it. The squeak is caused by the boards rubbing together, or rubbing against the fasteners (staples, nails, cleats, etc.).

A "crackling" floor is generally a sign of wood delamination due to excessive dryness. This means that, in the case of engineered hardwood, the layers of the plywood core are being stressed to the point of coming apart from each other. This may be mitigated by increasing the ambient humidity in the room, assuming the delamination has not progressed too far. Most manufacturers specify 30% to 50% relative humidity. Buildings and building materials respond to atmospheric changes. Therefore, while securing the flooring to the subfloor may help reduce and mitigate the problem, some squeaking will likely reoccur when the weather changes, perhaps in a different part of the room. A foundation is the lowest and supporting layer of a structure. Foundations are generally divided into two categories: shallow foundations and deep foundations.

### Historic Foundation Types

***Figure:*** *The simplest foundation, a padstone. Latvian Ethnographic Open Air Museum.*

## Earthfast or Post in Ground Construction

Buildings and structures have a long history of being built with wood in contact with the ground. Post in ground construction may technically have no foundation.Timber pilings were used on soft or wet ground even below stone or masonry walls. In marine construction and bridge building a crisscross of timbers or steel beams in concrete is called grillage.

## Padstones

Perhaps the simplest foundation is the padstone, a single stone which both spreads the weight on the ground and raises the timber off of the ground. Staddle stones are a specific type of padstone.

## Stone Foundations

Dry laid stone foundations are common in many parts of the world.

## Modern Foundation Types

***Shallow Foundations:*** Shallow foundations, often called footings, are usually embedded about a metre or so into soil. One common type is the spread footing which consists of strips or pads of concrete (or other materials) which extend below the frost line and transfer the weight from walls and columns to the soil or bedrock.

Another common type of shallow foundation is the slab-on-grade foundation where the weight of the building is transferred to the soil through a concrete slab placed at the surface. Slab-on-grade foundations can be reinforced mat slabs, which range from 25 cm to several metres thick, depending on the size of the building, or post-tensioned slabs, which are typically at least 20 cm for houses, and thicker for heavier structures.

## Deep Foundations

A deep foundation is used to transfer the load of a structure down through the upper weak layer of topsoil to the stronger layer of subsoil below. There are different types of deep footings including impact driven piles, drilled shafts, caissons, helical piles, geo-piers and earth stabilized columns. The naming conventions for different types of footings vary between different engineers. Historically, piles were wood, later steel, reinforced concrete, and pre-tensioned concrete.

## Monopile Foundation

A monopile foundation is a type of deep foundation which uses a single, generally large-diameter, structural element embedded into the earth to support all the loads (weight, wind, etc.) of a large above-surface structure.

A large number of monopile foundations have been utilised in recent years for economically constructing fixed-bottomoffshore wind farms in shallow-water subsea locations.

For example, a single wind farmoff the coast of England went online in 2008 with over 100 turbines, each mounted on a 4.7-metre-diameter monopile footing in ocean depths up to 18 metres of water.

### Design

Foundations are designed to have an adequate load capacity with limited settlement by a geotechnical engineer, and the footing itself may be designed structurally by a structural engineer.

The primary design concerns are settlement and bearing capacity. When considering settlement, total settlement and differential settlement is normally considered. Differential settlement is when one part of a foundation settles more than another part. This can cause problems to the structure the foundation is supporting.

## Light Fixture

A light fixture (US English), light fitting (UK English), or luminaire is an electrical device used to create artificial light by use of an electric lamp. All light fixtures have a fixture body and a light socket to hold the lamp and allow for its replacement. Fixtures may also have a switch to control the light.

Fixtures require an electrical connection to a power source; permanent lighting may be directly wired, and moveable lamps have a plug.

Light fixtures may also have other features, such as reflectors for directing the light, an aperture (with or without a lens), an outer shell or housing for lamp alignment and protection, and an electrical ballast or power supply. A wide variety of special light fixtures are created for use in the automotive lighting industry, aerospace, marine and medicine.

Portable light fixtures are often called "lamps", as in table lamp or desk lamp. In technical terminology, the *lamp* is the light source, what is typically called the light bulb.

The term *luminaire* is recommended by the International Electrotechnical Commission (IEC) for technical use.

### History

Fixture manufacturing began soon after production of the incandescent light bulb. When practical uses of fluorescent lighting were realised after 1939, the three leading companies to produce various fixtures were Lightolier, Artcraft Fluorescent Lighting Corporation, and Globe Lighting in the United States.

### Fixture Types

Light fixtures are classified by how the fixture is installed, the light function or lamp type.

### Free-Standing or Portable

- Table lamp fixtures, standard lamp fixtures, and office task light luminaires.
  - o Balanced-arm lamp is a spot light with an adjustable arm such as anglepoise or Luxo L1.
  - o Gooseneck lamp
  - o Nightlight

### Fixed

- Recessed light — the protective housing is concealed behind a ceiling or wall, leaving only the fixture itself exposed. The ceiling-mounted version is often called a downlight.
  - o "Cans" with a variety of lamps — this term is jargon for inexpensive downlighting products that are recessed into the ceiling, or sometimes for uplights placed on the floor. The name comes from the shape of the housing. The term "pot lights" is often used in Canada and parts of the US.
  - o Cove light — recessed into the ceiling in a long box against a wall.
  - o Torch lamp, torchière, or floor lamp.
  - o Troffer — recessed fluorescent light fixtures, usually rectangular in shape to fit into a drop ceiling grid.
- Surface-mounted light — the finished housing is exposed, not flush with surface
  - o Chandelier
  - o Pendant light — suspended from the ceiling with a chain or pipe

- o Sconce — provide up or down lights; can be used to illuminate artwork, architectural details; commonly used in hallways or as an alternative to overhead lighting.
- o Track lighting fixture — individual fixtures ("track heads") can be positioned anywhere along the track, which provides electric power.
- o Under-cabinet light — mounted below kitchen wall cabinets
- o Ceiling fan - May sometimes have a light, often referred to as a light kit mounted to it.
- o Emergency lighting or exit sign — connected to a battery backup or to an electric circuit that has emergency power if the mains power fails
- o High- and low-bay lighting — typically used for general lighting for industrial buildings and often big-box stores
- o Strip lights or Industrial lighting — often long lines of fluorescent lamps used in a warehouse or factory

***Figure:*** *A decorative outdoor lamp at Leeds Town Hall.*

***Figure:*** *A garden solar lamp is an example of landscape lighting*

- Outdoor lighting and landscape lighting — used to illuminate walkways, parking lots, roadways, building exteriors and architectural details, gardens, and parks.
  - o High-mast, usually pole- or stanchion-mounted — for landscape, roadways, and parking lots
  - o Bollard — A type of architectural outdoor lighting that is a short, upright ground-mounted unit typically used to provide cutoff type illumination for egress lighting, to light walkways, steps, or other pathways.
  - o Solar lamp
  - o Street light
  - o Yard light

### Special-Purpose Lights

- Accent light — Any directional light which highlights an object or attracts attention to a particular area

- Background light — for use in video production
- Blacklight
- Christmas lights-also called fairy lights or twinkle lights and are often used at Christmas and other holidays for decoration.
- Flood light
- Safelight (for use in a darkroom)
- Safety lamp (for use in coal mines)
- Searchlight (for military and advertising use)
- Security lighting
- Step light
- Strobe light
- Traffic light
- Theatrical
    - Stage lighting instrument
    - Intelligent lighting
    - Followspot
- Wallwasher

### Lamp Types

Xenon arc lamp, Yablochkov candle

***Fluorescent:*** Fluorescent lamp, compact fluorescent lamp (CFL), Induction lamp, blacklight.

***Fuel Lamps:*** Betty lamp, butter lamp, carbide lamp, gas lighting, kerosene lamp, oil lamp, rush light, torch, candle, Limelight, gas mantle

***Safety Lamps:*** Davy lamp&Geordie lamp

- Gas-discharge lamp and high-intensity discharge lamp (HID)

Mercury-vapor lamp, Metal-halide (HMI, HQI, CDM), Sodium vapor or "high-pressure sodium"

- Neon sign
- Incandescent lamp

A-lamp, Parabolic aluminized reflector lamp (PAR), reflector lamp (R), bulged reflector lamp (BR) (refer to lamp shapes)

- Obsolete types: limelight, carbon button lamp, Mazda (light bulb), Nernst glower
- Novelty: Lava lamp

- Special purpose: heat lamp, Globar, gas mantle
- Halogen – special class of incandescent lamps
- Nuclear:self-powered lighting
- Plasma lamp
- Light-emitting diode (LED) (solid-state lighting)

### *Light-Fixture Controls*

- 3-way 2-circuit switch
- Dimmer
- Light switch (often part of the light socket or power cord on portable fixtures)
- Lighting control system
- Motion detector
- Timer
- Touch
- X10 (industry standard) systems

## Plumbing

Plumbing is the system of pipes, drains fittings, valves, valve assemblies, and devices installed in a building for the distribution of water for drinking, heating and washing, and the removal of waterborne wastes, and the skilled trade of working with pipes, tubing and plumbing fixtures in such systems.

A plumber is someone who installs or repairs piping systems, plumbing fixtures and equipment such as water heaters and backflow preventers. The plumbing industry is a basic and substantial part of every developed economy due to the need for clean water, and sanitary collection and transport of wastes. The word "plumbing" comes from the Latin *plumbum* for lead, as pipes were once made from lead.

Plumbing is usually distinguished from water supply and sewage systems, in that a plumbing system serves one building, while water and sewage systems serve a group of buildings.

### *History*

Plumbing originated during ancient civilizations such as the Greek, Roman, Persian, Indian, and Chinese cities as they developed public baths and needed to provide potable water and drainage of wastes, for larger numbers of people. Standardized earthen plumbing pipes with broad flanges making use of asphalt for preventing leakages appeared in the urban settlements of the Indus Valley Civilization by 2700 B.C. The Romans used lead pipe inscriptions to prevent water theft.

***Figure:*** *Roman lead pipe with a folded seam, at the Roman Baths in Bath, England*

Improvement in plumbing systems was very slow, with virtually no progress made from the time of the Roman system of aqueducts and lead pipes. Plumbing was extremely rare until the growth of modern densely-populated cities in the 1800s. During this period, public health authorities began pressing for better waste disposal systems to be installed, to prevent or control epidemics of disease. Earlier, the waste disposal system had merely consisted of collecting waste and dumping it on the ground or into a river. Eventually the development of separate, underground water and sewage systems eliminated open sewage ditches and cesspools.

Most large cities today pipe solid wastes to sewage treatment plants in order to separate and partially purify the water, before emptying into streams or other bodies of water. For potable water use, galvanized iron piping was commonplace in the United States from the late 1800s until around 1960. After that period, copper piping took over, first soft copper with flared fittings, then with rigid copper tubing utilising soldered fittings.

The use of lead for potable water declined sharply after World War II because of increased awareness of the dangers of lead poisoning. At this time, copper piping was introduced as a better and safer alternative to lead pipes.

### *Materials*

Water systems of ancient times relied on gravity for the supply of water, using pipes or channels usually made of clay, lead, bamboo, wood, or stone. Hollowed wooden logs wrapped in steel banding were used for plumbing pipes, particularly water mains. Logs were used for water distribution in England close to 500 years ago. US cities began using hollowed logs in the late 1700s through the 1800s.

Present-day water-supply systems use a network of high-pressure pumps, and pipes in buildings are now made of copper, brass, plastic (particularly cross-linked polyethylene called PEX, which is estimated to be used in 60% of single-family homes), or other nontoxic material. Due to its toxicity, lead has not been used in modern water-supply piping since the 1930s in the United States, although lead was used in plumbing solder for drinking water until it was banned in 1986. Drain and vent lines are made of plastic, steel, cast-iron, or lead.

The straight sections of plumbing systems are called "pipes" or "tubes". A pipe is typically formed via casting or welding, whereas a tube is made through extrusion. Pipe normally has thicker walls and may be threaded or welded, while tubing is thinner-walled and requires special joining techniques such as brazing, compression fitting, crimping, or for plastics, solvent welding. These joining techniques are discussed in more detail in the piping and plumbing fittings article.

### *Components*

In addition to lengths of pipe or tubing, pipe fittings are used in plumbing systems, such as valves, elbows, tees, and unions. Pipe and fittings are held in place with pipe hangers and strapping.

Plumbing fixtures are exchangeable devices using water that can be connected to a building's plumbing system. They are considered to be "fixtures", in that they are semi-permanent parts of buildings, not usually owned or maintained separately. Plumbing fixtures are seen by and designed for the end-users. Some examples of fixtures include water closets (also known as toilets), urinals, bidets, showers, bathtubs, utility and kitchen sinks, drinking fountains, ice makers, humidifiers, air washers, fountains, and eye wash stations.

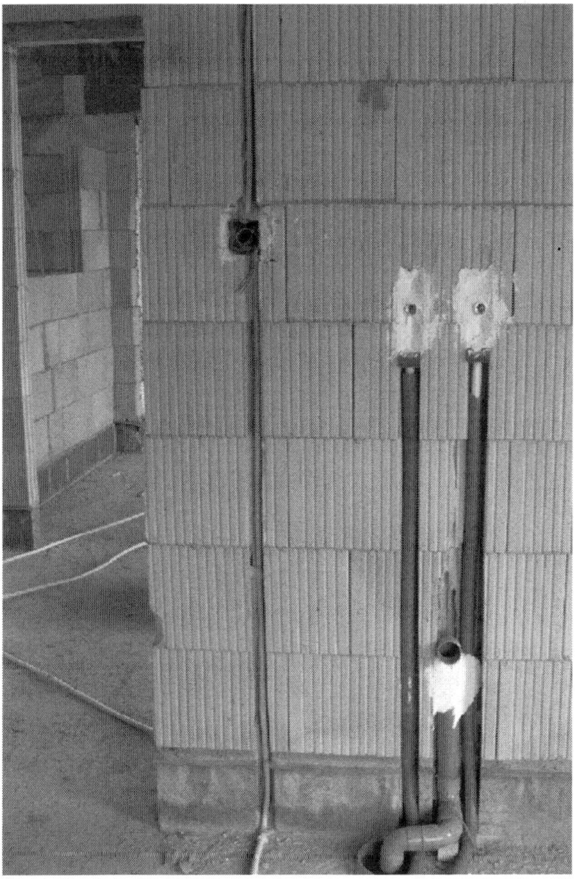

***Figure:*** *Piping being placed for a sink*

### Sealants

Threaded pipe joints are sealed with thread seal tape or pipe dope. Many plumbing fixtures are sealed to their mounting surfaces with plumber's putty.

### Equipment and Tools

Plumbing equipment includes devices often hidden behind walls or in utility spaces which are not seen by the general public. It includes water metres, pumps, expansion tanks, backflow arrestors, water filters, UV sterilization lights, water softeners, water heaters, heat exchangers, gauges, and control systems.

Specialised plumbing tools include pipe wrenches, flaring pliers, and joining tools such as soldering torches and crimp tools. New tools have been developed to help plumbers fix problems more efficiently. For example, plumbers use video cameras for inspections of hidden leaks or problems, they use hydro jets, and high pressure hydraulic pumps connected to steel cables for trench-less sewer line replacement.

### Systems

The major categories of plumbing systems or subsystems are:

- Potable cold and hot tap water supply
- Plumbing drainage venting
- Septic systems
- Rainwater, surface, and subsurface water drainage
- Fuel gas piping

For their environmental benefit and sizable energy savings, hot water heat recycling units are being installed in residential buildings. Ecological concerns and water shortages have increased interest in graywater recovery and treatment systems.

Plumbing may also include hydronics, which involves heating and cooling systems utilising water to transport thermal energy. The New York City steam system is an example of a large district heating system.

### Firestopping

Fire protection systems are largely used in commercial and residential buildings, and they may be a very simple system or an integrated solution which, beyond fire sensors and fire extinguishers methods based on spraying or sprinkling fire-suppressing fluids, includes the use of materials that have high fire-resistance coefficients to delay fire propagation. These materials and their use are regulated by specific standards that, not only manufacturers use in the fabrication process (flame retardant electric cables, for example), but also service providers in systems implementations and assemblies in the field.

Sometimes, a tube or another mechanical part has to penetrate a wall or floor that has special fire insulation material or that have special fire insulation resistance, it is needed to recover the original level of security of the fire system in case of maintenance, or to reach the designed level of security, so to avoid or delay the propagation of flame, smoke and heat accordingly.

Firestopping is a technique used in these cases, and it is required where mechanical penetrants traverse fire-resistance rated wall and floor assemblies, or membranes thereof. This work is usually done worldwide by the insulation trade and/or speciality firestop sub-contractors.

### Regulation

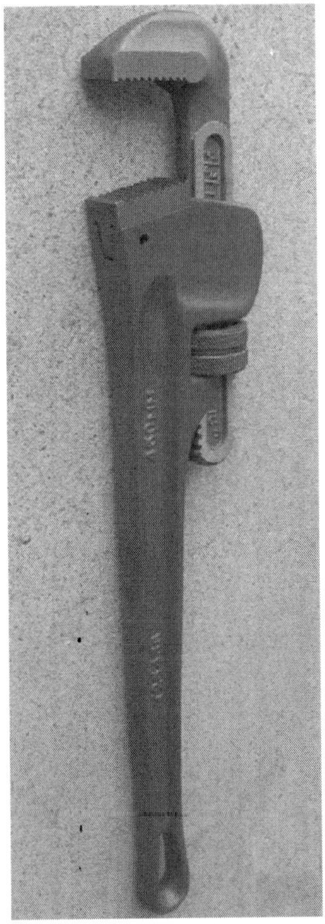

***Figure:*** *A pipe wrench for holding and turning pipe.*

Much of the plumbing work in populated areas is regulated by government or quasi-government agencies due to the direct impact on the public's health, safety, and welfare. Plumbing installation and repair work on residences and other buildings generally must be done according to plumbing and building codes to protect the inhabitants of the buildings and to ensure safe, quality construction to future buyers. If permits are required for work, plumbing contractors typically secure them from the authorities on behalf of home or building owners.

In the United Kingdom the professional body is the Chartered Institute of Plumbing and Heating Engineering (educational charity status) and it is true that the trade still remains virtually ungoverned; there are no systems in place to monitor or control the activities of unqualified plumbers or those home owners who choose to undertake installation and maintenance works themselves, despite the health and safety issues which arise from such works when they are undertaken incorrectly; see *Health Aspects of Plumbing (HAP)* published jointly by the World Health Organisation (WHO) and the World Plumbing Council (WPC). WPC has subsequently appointed a representative to the World Health Organisation to take forward various projects related to Health Aspects of Plumbing.

In the United States, plumbing codes and licensing are generally controlled by state and local governments. At the national level, the Environmental Protection Agency has set guidelines about what constitutes lead-free plumbing fittings and pipes, in order to comply with the Safe Drinking Water Act.

Some widely used Standards in the United States are:

- ASME A112.6.3 – Floor and Trench Drains
- ASME A112.6.4 – Roof, Deck, and Balcony Drains
- ASME A112.18.1/CSA B125.1 – Plumbing Supply Fittings
- ASME A112.19.1/CSA B45.2 – Enameled Cast Iron and Enameled Steel Plumbing Fixtures
- ASME A112.19.2/CSA B45.1 – Ceramic Plumbing Fixtures

## Roof

A roof is the covering on the uppermost part of a building. A roof protects the building and its contents from the effects of weather and the invasion of animals. Structures that require roofs range from a letter box to a cathedral or stadium, dwellings being the most numerous.

In most countries a roof protects primarily against rain. In Persia the citizens used their roofs for milling wheat, farming, gardens and extra space. Depending upon the nature of the building, the roof may also protect against heat, sunlight, cold, snow and wind. Other types of structure, for example, a garden conservatory, might use roofing that protects against cold, wind and rain but admits light. A verandah may be roofed with material that protects against sunlight but admits the other elements. The characteristics of a roof are dependent upon the purpose of the building that it covers, the available roofing materials and the local traditions of construction and wider concepts of architectural design and practice and may also be governed by local or national legislation.

### Design Elements

The elements in the design of a roof are:

- the material
- the construction
- the durability

The material of a roof may range from banana leaves, wheaten straw or seagrass to lamininated glass, copper, aluminium sheeting and precast concrete. In many parts of the world ceramictiles have been the predominant roofing material for centuries.

The construction of a roof is determined by its method of support and how the underneath space is bridged and whether or not the roof is *pitched*. The *pitch* is the angle at which the roof rises from its lowest to highest point.

Most US domestic architecture, except in very dry regions, has roofs that are sloped, or *pitched*. Although modern construction elements such as drainpipes may remove the need for pitch, roofs are pitched for reasons of tradition and aesthetics. So the pitch is partly dependent upon stylistic factors, and partially to do with practicalities.

Some types of roofing, for example thatch, require a steep pitch in order to be waterproof and durable. Other types of roofing, for example pantiles, are unstable on a steeply pitched roof but provide excellent weather protection at a relatively low angle. In regions where there is little rain, an almost flat roof with a slight run-off provides adequate protection against an occasional downpour. Drainpipes also remove the need for a sloping roof.

The durability of a roof is a matter of concern because the roof is often the least accessible part of a building for purposes of repair and renewal, while its damage or destruction can have serious effects.

### Form of a Roof

The shape of roofs differs greatly from region to region. The main factors which influence the shape of roofs are the climate and the materials available for roof structure and the outer covering.

The basic shapes of roofs are flat, skillion, gabled, hipped, arched and domed. There are many variations on these types. Roofs constructed of flat sections that are sloped are referred to as pitched roofs (generally if the angle exceeds 10 degrees). Pitched roofs, including gabled, hipped and skillion roofs, make up the greatest number of domestic roofs. Some roofs follow organic shapes, either by architectural design or because a flexible material such as thatch has been used in the construction.

### Parts of a Roof

There are two parts to a roof, its supporting structure and its outer skin, or uppermost weatherproof layer. In a minority of buildings, the outer layer is also a self-supporting structure. The roof structure is generally supported upon walls, although some building styles, for example, geodesic and A-frame, blur the distinction between wall and roof.

### Support

The supporting structure of a roof usually comprises beams that are long and of strong, fairly rigid material such as timber, and since the mid-19th century, cast iron or steel. In countries that use bamboo extensively, the flexibility of the material causes a distinctive curving line to the roof, characteristic of Oriental architecture. Timber lends itself to a great variety of roof shapes. The timber structure can fulfil an aesthetic as well as practical function, when left exposed to view.

Stone lintels have been used to support roofs since prehistoric times, but cannot bridge large distances. The stone arch came into extensive use in the ancient Roman period and in variant forms could be used to span spaces up to 140 feet (43 m) across. The stone arch or vault, with or without ribs, dominated the roof structures of major architectural works for about 2,000 years, only giving way to iron beams with the Industrial Revolution and the designing of such buildings as Paxton's Crystal Palace, completed 1851.

With continual improvements in steel girders, these became the major structural support for large roofs, and eventually for ordinary houses as well. Another form of girder is the reinforced concrete beam, in which metal rods are encased in concrete, giving it greater strength under tension.

### Outer Layer

This part of the roof shows great variation dependent upon availability of material. In simple vernacular architecture, roofing material is often vegetation, such as thatches, the most durable being sea grass with a life of perhaps 40 years. In many Asian countries bamboo is used both for the supporting structure and the outer layer where split bamboo stems are laid turned alternately and overlapped. In areas with an abundance of timber, wooden shingles are used, while in some countries the bark of certain trees can be peeled off in thick, heavy sheets and used for roofing.

The 20th century saw the manufacture of composition shingles which can last from a thin 20-year shingle to the thickest which are limited lifetime shingles, the cost depending on the thickness and durability of the shingle. When a layer of shingles wears out, they are usually stripped, along with the underlay and roofing nails, allowing a new layer to be installed. An alternative method is to install another layer directly over the worn layer. While this method is faster, it does not allow the roof sheathing to be inspected and water damage, often associated with worn shingles, to be repaired. Having multiple layers of old shingles under a new layer causes roofing nails to be located further from the sheathing, weakening their hold. The greatest concern with this method is that the weight of the extra material could exceed the dead load capacity of the roof structure and cause collapse.

Slate is an ideal, and durable material, while in the Swiss Alps roofs are made from huge slabs of stone, several inches thick. The slate roof is often considered the best type of roofing. A slate roof may last 75 to 150 years, and even longer. However, slate roofs are often expensive to install – in the USA, for example, a slate roof may have the same cost as the rest of the house. Often, the first part of a slate roof to fail is the fixing nails; they corrode, allowing the slates to slip. In the UK, this condition is known as "nail sickness". Because of this problem, fixing nails made of stainless steel or copper are recommended, and even these must be protected from the weather.

Asbestos, usually in bonded corrugated panels, has been used widely in the 20th century as an inexpensive, non-flammable roofing material with excellent insulating properties. Health and legal issues involved in the mining and handling of asbestos products means that it is no longer used as a new roofing material. However, many asbestos roofs continue to exist, particularly in South America and Asia.

Roofs made of cut turf (modern ones known as Green roofs, traditional ones as sod roofs) have good insulating properties and are increasingly encouraged as a way of "greening" the Earth. Adobe roofs are roofs of clay, mixed with binding material such as straw or animal hair, and plastered on lathes to form a flat or gently sloped roof, usually in areas of low rainfall.

In areas where clay is plentiful, roofs of baked tiles have been the major form of roof. The casting and firing of roof tiles is an industry that is often associated with brickworks. While the shape and colour of tiles was once regionally distinctive, now tiles of many shapes and colours are produced commercially, to suit the taste and pocketbook of the purchaser.

Sheet metal in the form of copper and lead has also been used for many hundreds of years. Both are expensive but durable, the vast copper roof of Chartres Cathedral, oxidised to a pale green colour, having been in place for hundreds of years. Lead, which is sometimes used for church roofs, was most commonly used as flashing in valleys and around chimneys on domestic roofs, particularly those of slate. Copper was used for the same purpose.

In the 19th century, iron, electroplated with zinc to improve its resistance to rust, became a light-weight, easily-transported, waterproofing material. Its low cost and easy application made it the most accessible commercial roofing, world wide. Since then, many types of metal roofing have been developed. Steel shingle or standing-seam roofs last about 50 years or more depending on both the method of installation and the moisture barrier (underlayment) used and are between the cost of shingle roofs and slate roofs. In the 20th century a large number of roofing materials were developed, including roofs based on bitumen (already used in previous centuries), on rubber and on a range of synthetics such as thermoplastic and on fibreglass.

### Functions of a Roof

***Insulation:*** Because the purpose of a roof is to protect people and their possessions from climatic elements, the insulating properties of a roof are a consideration in its structure and the choice of roofing material.

Some roofing materials, particularly those of natural fibrous material, such as thatch, have excellent insulating properties. For those that do not, extra insulation is often installed under the outer layer. In developed countries, the majority of dwellings have a ceiling installed under the structural members of the roof. The purpose of a ceiling is to insulate against heat and cold, noise, dirt and often from the droppings and lice of birds who frequently choose roofs as nesting places.

Concrete tiles can be used as insulation. When installed leaving a space between the tiles and the roof surface, it can reduce heating caused by the sun.

Forms of insulation are felt or plastic sheeting, sometimes with a reflective surface, installed directly below the tiles or other material; synthetic foam batting laid above the ceiling and recycled paper products and other such materials that can be inserted or sprayed into roof cavities. So called Cool roofs are becoming increasingly popular, and in some cases are mandated by local codes. Cool roofs are defined as roofs with both high reflectivity and high thermal emittance.

Poorly insulated and ventilated roofing can suffer from problems such as the formation of ice dams around the overhanging eaves in cold weather, causing water from melted snow on upper parts of the roof to penetrate the roofing material. Ice dams occur when heat escapes through the uppermost part of the roof, and the snow at those points melts, refreezing as it drips along the shingles, and collecting in the form of ice at the lower points. This can result in structural damage from stress, including the destruction of gutter and drainage systems.

### Drainage

The primary job of most roofs is to keep out water. The large area of a roof repels a lot of water, which must be directed in some suitable way, so that it does not cause damage or inconvenience. Flat roof of adobe dwellings generally have a very slight slope. In a Middle Eastern country, where the roof may be used for recreation, it is often walled, and drainage holes must be provided to stop water from pooling and seeping through the porous roofing material.

Similar problems, although on a very much larger scale, confront the builders of modern commercial properties which often have flat roofs. Because of the very large nature of such roofs, it is essential that the outer skin is of a highly impermeable material. Most industrial and commercial structures have conventional roofs of low pitch.

In general, the pitch of the roof is proportional to the amount of precipitation. Houses in areas of low rainfall frequently have roofs of low pitch while those in areas of high rainfall and snow, have steep roofs. The longhouses of Papua New Guinea, for example, being roof-dominated architecture, the high roofs sweeping almost to the ground. The high steeply-pitched roofs of Germany and Holland are typical in regions of snowfall. In parts of North America such as Buffalo, USA or Montreal, Canada, there is a required minimum slope of 6 inches in 12 inches, a pitch of 30 degrees. There are regional building styles which contradict this trend, the stone roofs of the Alpine chalets being usually of gentler incline. These buildings tend to accumulate a large amount of snow on them, which is seen as a factor in their insulation. The pitch of the roof is in part determined by the roofing material available, a pitch of 3/12 or greater slope generally being covered with asphalt shingles, wood shake, corrugated steel, slate or tile.

The water repelled by the roof during a rainstorm is potentially damaging to the building that the roof protects. If it runs down the walls, it may seep into the mortar or through panels. If it lies around the foundations it may cause seepage to the interior, rising damp or dry rot. For this reason most buildings have a system in place to protect the walls of a building from most of the roof water. Overhanging eaves are commonly employed for this purpose. Most modern roofs and many old ones have systems of valleys, gutters, waterspouts, waterheads and drainpipes to remove the water from the vicinity of the building. In many parts of the world, roofwater is collected and stored for domestic use.

Areas prone to heavy snow benefit from a metal roof because their smooth surfaces shed the weight of snow more easily and resist the force of wind better than a wood shingle or a concrete tile roof.

### Solar Roofs

Newer systems include solar shingles which generate electricity as well as cover the roof. There are also solar systems available that generate hot water or hot air and which can also act as a roof covering. More complex systems may carry out all of these functions: generate electricity, recover thermal energy, and also act as a roof covering.

Solar systems can be integrated with roofs by:

- integration in the covering of pitched roofs, e.g. solar shingles.
- mounting on an existing roof, e.g. solar panel on a tile roof.
- integration in a flat roof membrane using heat welding, e.g. PVC.
- mounting on a flat roof with a construction and additional weight to prevent uplift from wind.

## Stairs

A stairway, staircase, stairwell, flight of stairs, or simply stairs is a construction designed to bridge a large vertical distance by dividing it into smaller vertical distances, called steps. Stairs may be straight, round, or may consist of two or more straight pieces connected at angles.

Special types of stairs include escalators and ladders. Some alternatives to stairs are elevators (UK: lifts), stairlifts and inclined moving walkways as well as stationary inclined sidewalks (UK: pavements).

## Components and Terminology

***Step:*** The step is composed of the tread and riser.

***Tread:*** The part of the stairway that is stepped on. It is constructed to the same specifications (thickness) as any other flooring. The tread "depth" is measured from the outer edge of the step to the vertical "riser" between steps. The "width" is measured from one side to the other.

***Riser:*** The vertical portion between each tread on the stair. This may be missing for an "open" stair effect.

***Nosing:*** An edge part of the tread that protrudes over the riser beneath. If it is present, this means that, measured horizontally, the total "run" length of the stairs is not simply the sum of the tread lengths, as the treads actually overlap each other slightly.

***Starting Step or Bullnose:*** Where stairs are open on one or both sides, the first step above the lower floor may be wider than the other steps and rounded. The balusters typically form a semicircle around the circumference of the rounded portion and the handrail has a horizontal spiral called a "volute" that supports the top of the balusters.

Besides the cosmetic appeal, starting steps allow the balusters to form a wider, more stable base for the end of the handrail. Handrails that simply end at a post at the foot of the stairs can be less sturdy, even with a thick post. A double bullnose can be used when both sides of the stairs are open.

***Stringer, Stringer Board or Sometimes Just String:*** The structural member that supports the treads and risers. There are typically two stringers, one on either side of the stairs; though the treads may be supported many other ways. The stringers are sometimes notched so that the risers and treads fit into them. Stringers on open-sided stairs are often open themselves so that the treads are visible from the side. Such stringers are called "cut" stringers. Stringers on a closed side of the stairs are closed, with the support for the treads routed into the stringer.

***Winders:*** Winders are steps that are narrower on one side than the other. They are used to change the direction of the stairs without landings. A series of winders form a circular or spiral stairway.

When three steps are used to turn a 90° corner, the middle step is called a kite winder as a kite-shaped quadrilateral.

***Trim:*** Trim (e.g. quarter-round or baseboard trim) is normally applied where walls meet floors and often underneath treads to hide the reveal where the tread and riser meet.

Shoe moulding may be used between where the lower floor and the first riser meet.

Trimming a starting step is a special challenge as the last riser above the lower floor is rounded.

Flexible, plastic trim is available for this purpose, however wooden mouldings are still used and are either cut from a single piece of rounded wood, or bent with laminations.

Scotia is concave moulding that is underneath the nosing between the riser and the tread above it.

### The Railing System

***Figure:*** *A stairway with handrails leading to a panoramic tower of Harjun Vesilinna in Jyväskylä, Finland.*

***Figure:*** *A part of stairs connecting between 2nd and 3rd floor.*

***Figure:*** *A Stairway with a landing in the middle.*

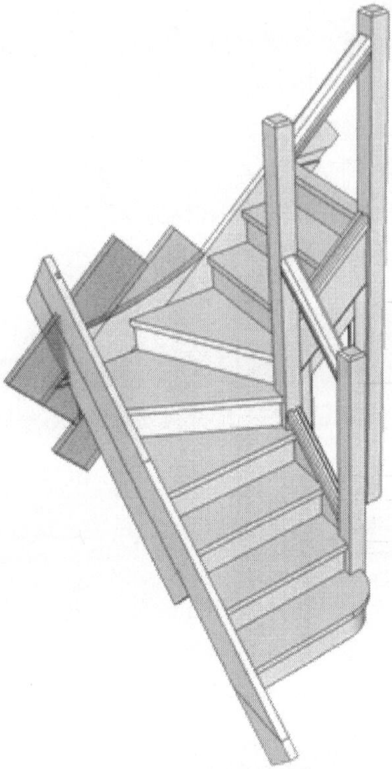

***Figure:*** *Example of Winder Stairs with a simple handrail supported by three newel posts.*

The balustrade is the system of railings and balusters that prevents people from falling over the edge.

***Banister, Railing or Handrail:*** The angled member for handholding, as distinguished from the vertical balusters which hold it up for stairs that are open on one side; there is often a railing on both sides, sometimes only on one side or not at all, on wide staircases there is sometimes also one in the middle, or even more. The term "banister" is sometimes used to mean just the handrail, or sometimes the handrail and the balusters or sometimes just the balusters.

***Volute:*** A handrail end element for the bullnose step that curves inward like a spiral. A volute is said to be right or left-handed depending on which side of the stairs the handrail is as one faces up the stairs.

***Turnout:*** Instead of a complete spiral volute, a turnout is a quarter-turn rounded end to the handrail.

***Gooseneck:*** The vertical handrail that joins a sloped handrail to a higher handrail on the balcony or landing is a gooseneck.

***Rosette:*** Where the handrail ends in the wall and a half-newel is not used, it may be trimmed by a rosette.

***Easings:*** Wall handrails are mounted directly onto the wall with *wall brackets*. At the bottom of the stairs such railings flare to a horizontal railing and this horizontal portion is called a "starting easing". At the top of the stairs, the horizontal portion of the railing is called a "over easing".

***Core Rail:*** Wood handrails often have a metal core to provide extra strength and stiffness, especially when the rail has to curve against the grain of the wood. The archaic term for the metal core is "core rail".

***Baluster:*** A term for the vertical posts that hold up the handrail. Sometimes simply called *guards* or *spindles*. Treads often require two balusters. The second baluster is closer to the riser and is taller than the first. The extra height in the second baluster is typically in the middle between decorative elements on the baluster. That way the bottom decorative elements are aligned with the tread and the top elements are aligned with the railing angle.

***Newel:*** A large baluster or post used to anchor the handrail. Since it is a structural element, it extends below the floor and subfloor to the bottom of the floor joists and is bolted right to the floor joist. A *half-newel* may be used where a railing ends in the wall. Visually, it looks like half the newel is embedded in the wall. For open landings, a newel may extend below the landing for a decorative *newel drop*.

***Finial:*** A decorative cap to the top of a newel post, particularly at the end of the balustrade.

***Baserail or Shoerail:*** For systems where the baluster does not start at the treads, they go to a baserail. This allows for identical balusters, avoiding the second baluster problem.

***Fillet:*** A decorative filler piece on the floor between balusters on a balcony railing.

Handrails may be *continuous* (sometimes called *over-the-post*) or *post-to-post* (or more accurately "newel-to-newel"). For continuous handrails on long balconies, there may be multiple newels and *tandem caps* to cover the newels. At corners, there are *quarter-turn caps*. For post-to-post systems, the newels project above the handrails.

Another, more classical, form of handrailing which is still in use is the tangent method. A variant of the Cylindric method of layout, it allows for continuous climbing and twisting rails and easings. It was defined from principles set down by architect Peter Nicholson in the 18th century.

### Other Terminology

***Figure:*** *Historical photo of a staircase in the Ford plant in Los Angeles with a double bullnose and two volutes. The photo also shows an intermediate landing as part of this U-shaped stair.*

***Balcony:*** For stairs with an open concept upper floor or landing, the upper floor is functionally a balcony. For a straight flight of stairs, the balcony may be long enough to require multiple newels to support the length of railing. In modern homes, it is common to have hardwood floors on the first floor and carpet on the second. The homeowner should consider using hardwood nosing in place of carpet. Should the carpet be subsequently replaced with hardwood, the balcony balustrade may have to be removed to add the nosing.

***Flight:*** A flight is an uninterrupted series of steps.

***Floating Stairs:*** A flight of stairs is said to be "floating" if there is nothing underneath. The risers are typically missing as well to emphasize the open effect. There may be only one stringer or the stringers otherwise minimized. Where building codes allow, there may not even be handrails.

***Landing or Platform:*** A landing is the area of a floor near the top or bottom step of a stair. An intermediate landing is a small platform that is built as part of the stair between main floor levels and is typically used to allow stairs to change directions, or to allow the user a rest.

A half landing is where a 180° change in direction is made, and a quarter landing is where a 90° change in direction is made (on an intermediate landing). As intermediate landings consume floor space they can be expensive to build. However, changing the direction of the stairs allows stairs to fit where they would not otherwise, or provides privacy to the upper level as visitors downstairs cannot simply look up the stairs to the upper level due to the change in direction.

***Mobile Safety Steps:*** Can be used as temporary, safe replacements for many types of stairs

***Runner:*** Carpeting that runs down the middle of the stairs. Runners may be directly stapled or nailed to the stairs, or may be secured by a specialised bar that holds the carpet in place where the tread meets the riser, known as a stair rod.

***Spandrel:*** If there is not another flight of stairs immediately underneath, the triangular space underneath the stairs is called a "spandrel". It is frequently used as a closet.

***Staircase:*** This term is often reserved for the stairs themselves: the steps, railings and landings; though often it is used interchangeably with "stairs" and "stairway". In the UK, however, the term "staircase" denotes what in the U.S. is called "stairway", but usually includes the casing – the walls, bannisters and underside of the stairs or roof above.

***Stairway:*** This primarily American term is often reserved for the entire stairwell and staircase in combination; though often it is used interchangeably with "stairs" and "staircase".

***Stairwell:*** The spatial opening, usually a vertical shaft, containing an indoor stairway; by extension it is often used as including the stairs it contains.

### *Measurements*

The measurements of a stair, in particular the rise height and going of the steps, should remain the same along the stairs.

The following stair measurements are important:

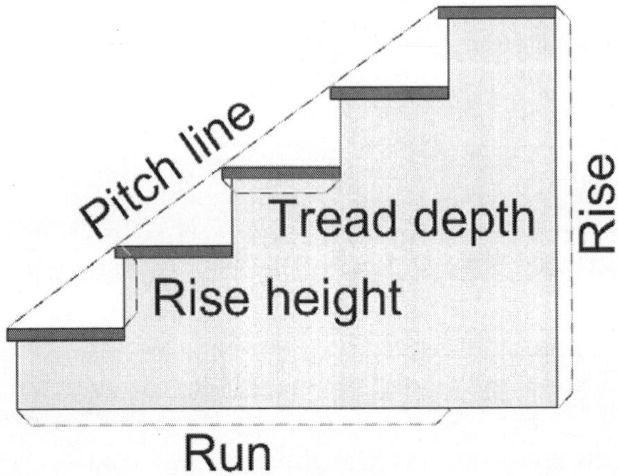


- The rise height or rise of each step is measured from the top of one tread to the next. It is not the physical height of the riser; the latter excludes the thickness of the tread. A person using the stairs would move this distance vertically for each step he takes.
- The tread depth of a step is measured from the edge of the nosing to the vertical riser; if the steps have no nosing, it is the same as the going; otherwise it is the going plus the extent of one nosing.
- The going of a step is measured from the edge of the nosing to the edge of nosing in plan view. A person using the stairs would move this distance forward with each step they take.
- To avoid confusion, the number of steps in a set of stairs is always the number of risers, not the number of treads.
- The total run or total going of the stairs is the horizontal distance from the first riser to the last riser. It is often not simply the sum of the individual tread lengths due to the nosing overlapping between treads. If there are $N$ steps, the total run equals $N$-1 times the going: the tread of the last step is part of a landing and is not counted.
- The total rise of the stairs is the height between floors (or landings) that the flight of stairs is spanning. If there are $N$ steps, the total rise equals $N$ times the rise of each step.

- The slope or pitch of the stairs is the ratio between the rise and the going (not the tread depth, due to the nosing). It is sometimes called the rake of the stairs. The pitch line is the imaginary line along the tip of the nosing of the treads. In the UK, stair pitch is the angle the pitch line makes with the horizontal, measured in degrees. The value of the slope, as a ratio, is then the tangent of the pitch angle.
- Headroom is the height above the nosing of a tread to the ceiling above it.
- Walkline – for curved stairs, the inner radius of the curve may result in very narrow treads. The "walkline" is the imaginary line some distance away from the inner edge on which people are expected to walk. Building code will specify the distance. Building codes will then specify the minimum tread size at the walkline.

The easiest way to calculate the rise and run is to use a stair stringer calculator.

### Ergonomics and Building Code Requirements

Ergonomically and for safety reasons, stairs must have certain measurements so that people can comfortably use them. Building codes typically specify certain measurements so that the stairs are not too steep or narrow. American building codes, while varying from State to State and County to County, generally specify the following parameters:

- Minimum tread length, typically 9 inches (229 mm) excluding the nosing for private residences. Some building codes also specify a minimum riser height, often 5 inches (127 mm).
- Riser-Tread formula: Sometimes the stair parameters will be something like riser + tread equals 17–18 inches (432–457 mm); another formula is 2 times riser + tread equals 24.6 inches (625 mm), the length of a stride. Thus a 7 inches (178 mm) rise and a 10.6 inches (269 mm) tread exactly meets this code. If only a 2 inches (51 mm) rise is used then a 20.6 inches (523 mm) tread is required. This is based on the principle that a low rise is more like walking up a gentle incline and so the natural swing of the leg will be longer.
- Low rise stairs are very expensive in terms of the space consumed. Such low rise stairs were built into the Winchester Mystery House to accommodate the infirmities of the owner,

Sarah Winchester, before the invention of the elevator. These stairways, called "Easy Risers" consist of five flights wrapped into a multi-turn arrangement with a total width equal to more than four times the individual flight width and a depth roughly equal to one flight's run plus this width. The flights have varying numbers of steps.

- Slope: A value for the rise-to-tread ratio of 17/29 H" 0.59 is considered optimal; this corresponds to a pitch angle of about 30°.
- Variance on riser height and tread depth between steps on the same flight should be very low. Building codes require variances no larger than 0.1875 inches (4.76 mm) between depth of adjacent treads or the height of adjacent risers; within a flight, the tolerance between the largest and smallest riser or between the largest and smallest tread can not exceed 0.375 inches (9.5 mm). The reason is that on a continuous flight of stairs, people get used to a regular step and may trip if there is a step that is different, especially at night. The general rule is that all steps on the same flight must be identical. Hence, stairs are typically custom made to fit the particular floor to floor height and horizontal space available. Special care must be taken on the first and last risers. Stairs must be supported directly by the subfloor. If thick flooring (e.g. thick hardwood planks) are added on top of the subfloor, it will cover part of the first riser, reducing the effective height of the first step. Likewise at the top step, if the top riser simply reaches the subfloor and thick flooring is added, the last rise at the top may be higher than the last riser. The first and last riser heights of the rough stairs are modified to adjust for the addition of the finished floor.
- Maximum nosing protrusion, typically 1.25 inches (32 mm) to prevent people from tripping on the nosing.
- Height of the handrail. This is typically between 34 and 38 inches (864 and 965 mm), measured to the nose of the tread. The minimum height of the handrail for landings may be different and is typically 36 inches (914 mm).
- Handrail diameter. The size has to be comfortable for grasping and is typically between 1.25 and 2.675 inches (32 and 67.9 mm).
- Maximum space between the balusters of the handrail. This is typically 4 inches (102 mm).

- Openings (if they exist) between the bottom rail and treads are typically no bigger than 6 inches (152 mm).
- Headroom: At least 83 inches (211 cm).
- Maximum vertical height between floors or landings. This allows people to rest and limits the height of a fall.
- Mandate handrails if there is more than a certain number of steps (typically 2 risers)
- Minimum width of the stairway, with and without handrails
- Not allow doors to swing over steps; the arc of doors must be completely on the landing/floor.
- A stairwell may be designated as an area of refuge as well as a fire escape route, due to its fire-resistance rated design and fresh air supply.

Nicolas-François Blondel in the last volume of his Coursd' architecture (1675-1683) was the first known person to establish the ergonomic relationship of tread and riser dimensions. He specified that 2 x riser + tread = step length.

It is estimated that a noticeable mis-step occurs once in 7,398 uses and a minor accident on a flight of stairs occurs once in 63,000 uses. Some people choose to live in residences without stairs so that they are protected from injury.

Stairs are not suitable for wheelchairs and other vehicles. A stairlift is a mechanical device for lifting wheelchairs up and down stairs. For sufficiently wide stairs, a rail is mounted to the treads of the stairs, or attached to the wall. A chair is attached to the rail and the person on the chair is lifted as the chair moves along the rail.

## USA Building Regulations

***UK Building Regulations:*** Building regulations are required for stairs used where the difference of level is greater than 600 mm (23.6 in)

Steepness of stairs – Rise and Going - Any rise between 155 and 220 mm (6.1 and 8.7 in) used with any going (tread) between 245 and 260 mm (9.6 and 10.2 in) or any rise between 165 and 200 mm (6.5 and 7.9 in) used with any going between 223 and 300 mm (8.8 and 11.8 in)

Maximum Rise 220 mm (8.7 in) and Minimum Going 220 mm (8.7 in) remembering that the maximum pitch of private stairs is 42p . The normal relationship between dimensions of the rise and going is that twice the rise plus the going (2R + G) should be between 550 and 700 mm (21.7 and 27.6 in)

Construction of steps - Steps should have level treads, they may have open risers but if so treads should overlap at least 16 mm (0.6 in). Domestic private stairs are likely to be used by children under 5 years old so they should be constructed so that a 100 mm (3.9 in) diameter sphere cannot pass though the opening in the risers.

Headroom -A headroom of 2 m (78.7 in) is adequate. Special considerations can be made for loft conversions see paragraph 1.10 of the approved document for clarification on the 1.9 tapering to 1.8 m (70.9 in). Width of flights - No recommendations are given for stair widths in the Approved document K for stairs but designer's attention is drawn to approved document B: fire safety.

Length of flights - The approved document refers to 16 risers (steps) for stairs in shops or assembly building. There is no requirement for private stairs. In practice there will be less than 16 steps as 16 mm × 220 mm (0.6 in × 8.7 in) gives over 3.5 m (137.8 in) total rise which is way above that in a domestic situation.

Landings - Level, unobstructed landings should be provided at the top and bottom of every flight. The width and length being at least that of the width of the stairs and can include part of the floor. A door may swing across the landing at the bottom of the flight but must leave a clear space of at least 400 mm (15.7 in) across the whole landing.

Tapered steps - There are special rules for stairs with tapered steps as shown in the image Example of Winder Stairs above.

Alternate tread stairs can be provide in space saving situations

Guarding - Flights and landings must be guarded at the sides where the drop is more than 600 mm (23.6 in). As domestic private stairs are likely to be used by children under 5 the guarding must be constructed so that a 100 mm (3.9 in) diameter sphere cannot pass through any opening or constructed so that children will not be able to climb the guarding. The height for internal private stairs should be at least 900 mm (35.4 in) and be able to withstand a horizontal force of 0.36 kN/m (2.1 lbf/in).

### Forms

Stairs can take a large number of forms, combining winders and landings.

The simplest form is the straight flight of stairs, with neither winders nor landings. It is not often used in modern homes because:

- the upstairs is directly visible from the bottom of a straight flight of stairs.
- it is potentially more dangerous in that a fall is not interrupted until the bottom of the stairs.
- a straight flight requires enough space for the entire run of the stairs.

However, a straight flight of stairs is easier to design and construct than one with landings. Additionally, the rhythm of stepping is not interrupted in a straight run, which may offset the increased fall risk by helping to prevent a misstep in the first place.

"L" shaped stairways have one landing and usually change in direction by 90 degrees. "U" shaped stairs may employ a single wider landing for a change in direction of 180 degrees, or 2 landings for two changes in direction of 90 degrees each. Use of landings and a possible change of direction have the following effects:

- The upstairs is not directly visible from the bottom of the stairs, which can provide more privacy for the upper floor.
- A fall can be arrested at the landing.
- Though the landings consume total floor space, there is no requirement for a large single dimension, allowing more flexible floorplan designs.
- For larger stairs, particularly in exterior applications, a landing can provide a place to rest the legs.

A mono string staircase is a term used for a steel spine staircase with treads.

A Double string staircase has two steel beams on either side and treads in the centre.

### Spiral and Helical Stairs

Spiral stairs wind around a newel (also the central pole). They typically have a handrail on the outer side only, and on the inner side just the central pole. A squared spiral stair assumes a square stairwell and expands the steps and railing to a square, resulting in unequal steps (larger where they extend into a corner of the square). A pure spiral assumes a circular stairwell and the steps and handrail are equal and positioned screw-symmetrically. A tight spiral stair with a central pole is very space efficient in the use of floor area. Spiral stairs have the disadvantage of being very steep. Unless the central column is very large, the circumference of the circle at the walk line will be small enough that it will be impossible to maintain a normal tread depth and a normal rise height without compromising headroom before reaching the upper floor. To maintain headroom most spiral stairs have very high rises and a very short going. Most building codes limit the use of spiral stairs to small areas or secondary usage.

The term "spiral" has a more narrow definition in a mathematical context, as a mathematical spiral lies in a single plane and moves towards or away from a central point. The mathematical term for motion where the locus remains at a fixed distance from a fixed line whilst moving in a circular motion about it is "helical". The presence or otherwise of a central pole does not affect the terminology applied to the design of the structure.

Spiral stairs in medieval times were generally made of stone and typically wound in a clockwise direction (from the ascender's point of view), to place attacking swordsmen (who were most often right-handed) at a disadvantage. This asymmetry forces the right-handed swordsman to engage the central pike and degrade his mobility compared with the defender who is facing down the stairs. Extant 14th to 17th century examples of these stairways can be seen at Muchalls Castle, Crathes Castle and Myres Castle in Scotland. Exceptions to the rule exist, however, as may be seen in the accompanying image of the Scala of the Palazzo Contarini del Bovolo, which winds up counter-clockwise.

Developments in manufacturing and design have led to the introduction of kit form spiral stairs. Steps and handrails can be bolted together to form a complete unit. These stairs can be made out of steel, timber, concrete or a combination of materials.

Helical or circular stairs do not have a central pole and there is a handrail on both sides. These have the advantage of a more uniform tread width when compared to the spiral staircase. Such stairs may also be built around an elliptical or oval planform. A double helix is possible, with two independent helical stairs in the same vertical space, allowing one person to ascend and another to descend, without ever meeting if they choose different helices (examples: Pozzo di S. Patrizio, Château de Chambord, Château de Blois, Crédit Lyonnais headquarters in Paris). Fire escapes, though built with landings and straight runs of stairs, are often functionally double helices, with two separate stairs intertwined and occupying the same floor space. This is often in support of legal requirements to have two separate fire escapes.

Both spiral and helical stairs can be characterized by the number of turns that are made. A "quarter-turn" stair deposits the person facing 90 degrees from the starting orientation. Likewise there are half-turn, three-quarters-turn and full-turn stairs. A continuous spiral may make many turns depending on the height. Very tall multi-turn spiral staircases are usually found in old stone towers within fortifications, churches and in lighthouses.

Winders may be used in combination with straight stairs to turn the direction of the stairs. This allows for a large number of permutations.

### History

The earliest spiral staircases appear in Temple A in the Greek colony Selinunte, Sicily, to both sides of the cella. The temple was constructed around 480–470 BC.

### Alternating Tread Stairs

Where there is insufficient space for the full run length of normal stairs, alternating tread stairs may be used. Alternating tread stairs allow for safe forward-facing descent of very steep stairs. The treads are designed such that they alternate between treads for each foot: one step is wide on the left side; the next step is wide on the right side. There is insufficient space on the narrow portion of the step for the other foot to stand, hence the person must always use the correct foot on the correct step. The slope of alternating tread stairs can be as high as 65 degrees as opposed to standard stairs which are almost always less than 45 degrees. The advantage of alternating tread stairs is that people can descend face forward. The only other alternative in such short spaces would be a ladder which requires backward-facing descent. Alternating tread stairs may not be safe for small children, the elderly or the physically challenged. Building codes typically classify them as ladders and will only allow them where ladders are allowed, usually basement or attic utility or storage areas not frequently accessed.

The image on the right illustrates the space efficiency gained by an alternating tread stair. The alternating tread stair appears in the image's centre, with green-coloured treads. The alternating stair requires one unit of space per step: the same as the half-width step on its left, and half as much as the full-width stair on its right. Thus, the horizontal distance between steps is in this case reduced by a factor of two, reducing the size of each step.

The horizontal distance between steps is reduced by a factor less than two if for construction reasons there are narrow "unused" steps. There is often (here also) glide plane symmetry: the mirror image with respect to the vertical centre plane corresponds to a shift by one step.

Alternating tread stairs have been in use since at least 1888.

### Notable Sets of Stairs

- The longest stairway is listed by Guinness Book of Records as the service stairway for the Niesenbahn funicular railway near Spiez, Switzerland, with 11,674 steps and a height of 1,669 m (5,476 ft). The stairs are employee-only.
- A flight of 7,200 steps (including inner temple Steps), with 6,293 Official Mountain Walkway Steps, leads up the East Peak of Mount Tai in China.
- The Ha»ikû Stairs, on the island of O»ahu, Hawai»i, are approximately 4,000 steps which climb nearly $^1/_2$ mile (0.8 km). Originally used to access longwire radio antennas which were strung high above the Ha»ikû Valley, between Honolulu and Kâne»ohe, they are closed to hikers.
- The Flørli stairs, in Lysefjorden, Norway, have 4,444 wooden steps which climb from sea level to 740 metres (2,428 feet). It is a maintenance stairway for the water pipeline to the old Flørli hydro plant. The hydro plant is now closed down, and the stairs are open to the public. The stairway is claimed to be the longest wooden stairway in the world.
- The CN Tower's staircase reaches the main deck level after 1,776 steps and the Sky Pod above after 2,579 steps; it is the tallest metal staircase on Earth.
- The Penrose stairs, devised by Lionel and Roger Penrose, are a famous impossible object. The image distorts perspective in such a manner that the stairs appear to be never-ending, a physical impossibility. The image was adopted by M. C. Escher in his iconic lithograph *Ascending and Descending.*
- The World Trade Centre Survivors' Staircase is the last visible structure above ground level at the World Trade Centre site. It was originally two outdoor flights of granite-clad stairs and an escalator that connected Vesey Street to the World Trade Centre's Austin J. Tobin Plaza. During the September 11,

2001, attacks, the stairs served as an escape route for hundreds of evacuees from 5 World Trade Centre, a 9-floor building adjacent to the 110-story towers.

- Stairwell A was the lone stairway left intact after the second plane hit the South Tower of the World Trade Centre during the September 11 attacks. It was believed to have remained intact until the South Tower collapsed at 9:59 am. 14 people were able to escape the floors located at the impact zone (including one man who saw the plane coming at him), and 4 people from the floors above the impact zone. Numerous 911 operators who received calls from individuals inside the South Tower were not well informed of the situation as it rapidly unfolded in the South Tower. Many operators told callers not to descend the tower on their own, even though it is now believed that Stairwell A was most likely passable at and above the point of impact.

# 5

# Wall Construction

A wall is a vertical structure, usually solid, that defines and sometimes protects an area. Most commonly, a wall delineates a building and supports its superstructure, separates space in buildings into sections, or protects or delineates a space in the open air. There are three principal types of structural walls: building walls, exterior boundary walls, and retaining walls.

***Figure:*** *Decorative exterior wall, Mexico City, Mexico, 2008.*

Building walls have one main purpose: to support roofs and ceilings. Such walls most often have three or more separate components.

In today's construction, a building wall will usually have the structural elements (such as 2×4 studs in a house wall), insulation, and finish elements or surface (such as drywall or panelling). In addition, the wall may house various types of electrical wiring or plumbing. Electrical outlets are usually mounted in walls.

Building walls frequently become works of art externally and internally, such as when featuring mosaic work or when murals are painted on them; or as design foci when they exhibit textures or painted finishes for effect. On a ship, the walls separating compartments are termed "bulkheads", whilst the thinner walls separating cabins are termed "partitions".

In architecture and civil engineering, the term curtain wall refers to the facade of a building which is not load-bearing but functions as decoration, finish, front, face, or history preservation.

## Type of Walls

### *Partition Wall*

A partition wall is a wall for the purpose of separating rooms, or dividing a room. Partition walls are usually not load-bearing. Partition walls may be constructed with bricks or blocks from clay, terra-cotta or concrete, reinforced, or hollow. Glass blocks may also be used.

They may also be constructed from sheet glass. Glass partition walls are a series of individual toughened glass panels, which are suspended from or slide along a robust aluminium ceiling track. The system does not require the use of a floor guide, which allows easy operation and an uninterrupted threshold.

Timber may be used. This type of partition consists of a wooden framework either supported on the floor below or by side walls. Metal lath and plaster, properly laid, forms a reinforced partition wall. Partition walls constructed from fibre cement sheeting are popular as bases for tiling in kitchens or in wet areas like bathrooms. Galvanized sheet fixed to wooden or steel members are mostly adopted in works of temporary character. Plain or reinforced partition walls may also be constructed from concrete, including pre-cast concrete blocks.

Metal framed partitioning is also available. This partition consists of track (used primarily at the base and head of the partition) and stud (vertical sections fixed at 600mm centres). Internal wall partitions also known as office partitioning is made using plasterboard (drywall), or varieties of glass. Toughened glass is a common option as it is feasible however there is also low iron glass better known as opti-white glass which increases light and solar heat transmission.

Wall partitions are constructed using beads and tracking which are either hung from the ceiling or fixed into the ground. The panels are inserted into the tracking and fixed.

There are variations of wall partitions which include the level of fire resistance they have, and their acoustic performance rating.

***Movable Partitions:*** Movable partitions are used where the walls of a room are frequently opened to form one large floor area. In this system, there are several types of partitions which include;

- Sliding: Sliding partitions consists of series of panels that slide in tracks fixed to the floor and ceiling. The machine of the partition is similar to those of sliding doors.
- Sliding & folding: Sliding and folding partitions operate in a similar manner to sliding folding doors. They are normally used for smaller spans.
- Screens: Screens are usually constructed of a metal or timber frame. It is fixed with plywood and chipboard inside. The screen supported with legs for free standing and easy movement.
- Pipe and drape: Fixed or telescopic uprights and horizontals provide a ground supported drape system with removable panels.

### Boundary Walls

Boundary walls include privacy walls, boundary-marking walls on property, and town walls. These intergrade into fences; the conventional differentiation is that a fence is of minimal thickness and often is open in nature, while a wall is usually more than a nominal thickness and is completely closed, or opaque. More to the point, if an exterior structure is made of wood or wire, it is generally referred to as a fence, while if it is made of masonry, it is considered a wall. A common term for both is barrier, convenient if it is partly a wall and partly a fence, for example the Berlin Wall. Another kind of wall/fence ambiguity is the ha-ha which is set below ground level, so as not to interrupt a view yet acting as a barrier to cattle for example.

Before the invention of artillery, many of the world's cities and towns, particularly in Europe and Asia, had defencive or protective walls (also called town walls or city walls). In fact, the English word "wall" is derived from Latin *vallum*, which was a type of fortification wall. Since they are no longer relevant for defence, such cities have grown beyond their walls, and many of the walls, or portions thereof, have been torn down, for example in Rome, Italy and in Beijing, China. Examples of protective walls on a much larger scale include the Great Wall of China and Hadrian's Wall.

### *Border Walls*

Some walls are designed to formally separate the border between one population and another. A border wall is constructed to limit the movement of people across a certain line or border. These structures vary in placement with regard to international borders and topography. The most famous example of border barrier in history is probably the Great Wall of China, a series of walls separating the Empire of China from nomadic powers to the north; the most prominent recent example was the Berlin Wall that surrounded the enclave of West Berlin, separating it from East Germany, during most of the Cold War era.

### *Retaining Walls*

In areas of rocky soils around the world, farmers have often pulled large quantities of stone out of their fields to make farming easier and have stacked those stones to make walls that either mark the field boundary, or the property boundary, or both. Retaining walls are a special type of wall, that may be either external to a building or part of a building, that serves to provide a barrier to the movement of earth, stone or water. The ground surface or water on one side of a retaining wall will be noticeably higher than on the other side. A dike is one type of retaining wall, as is a levee, a load-bearing foundation wall, and a sea wall.

### *Shared Walls*

Special laws often govern walls shared by neighbouring properties. Typically, one neighbour cannot alter the common wall if it is likely to affect the building or property on the other side. A wall may also separate apartment or hotel rooms from each other. Each wall has two sides and breaking a wall on one side will break the wall on the other side.

### *Portable Walls*

Portable walls, such as room dividers or portable partitions, are used to take a large open space and effectively divide it into smaller rooms. Portable walls can be static such as cubicle walls, or they can be wall panels mounted on casters to provide an easy way to reconfigure assembly space. They are often found inside schools, churches, convention centres, hotels and corporate facilities.

### Etymology

It is notable that English uses the same word to refer to an external wall, and the internal sides of a room. This is by no means universal, and many languages distinguish between the two. In German, some of this distinction can be seen between *Wand* and *Mauer*, in Spanish between *pared* and *muro*.

### Walls in Popular Culture

Walls are often seen in popular culture representing barriers preventing progress or entry. For example, the progressive/psychedelic rock band Pink Floyd used a metaphorical wall to represent the isolation felt by the protagonist of their 1979 concept album *The Wall*. American poet laureate Robert Frost describes a pointless rock wall as a metaphor for the myopia of the culture-bound in his poem "Mending Wall", published in 1914. In a real-life example, the Berlin Wall, constructed by the Soviet Union to divide Berlin into NATO and Warsaw Pact zones of occupation, became a worldwide symbol of oppression and isolation.

In some cases, a wall may refer to an individual's debilitating mental or physical condition, seen as an impassable barrier. Another common usage is as a communal surface to write upon. For instance the social networking site Facebook previously used an electronic "wall" to log the scrawls of friends until it was replaced by the "timeline" feature. In the popular 2013 anime Attack on Titan, walls were created to keep humans from their predator foes, the Titans. Later in the manga version it is revealed that the Titans may be the ones who made them.

## Door

A door is an opening or closing structure used to block off an entrance, typically consisting of an interior side that faces the inside of a space and an exterior side that faces the outside of that space. While in some cases the interior side of a door may match its exterior side, in other cases there are sharp contrasts between the two sides, such as in the case of the vehicle door. In addition, doors typically consist of a panel that swings on hinges or that slides or spins inside of a space.

When open, doors admit people, animals, ventilation, and light. The door is used to control the physical atmosphere within a space by enclosing the air drafts, so that interiors may be more effectively heated or cooled. Doors are significant in preventing the spread of fire.

They act as a barrier to noise. Many doors are equipped with locking mechanisms to allow entrance to certain people and keep out others.

Doors are used to screen areas of a building for aesthetics, keeping formal and utility areas separate. Doors also have an aesthetic role in creating an impression of what lies beyond. Doors are often symbolically endowed with ritual purposes, and the guarding or receiving of the keys to a door, or being granted access to a door can have special significance. Similarly, doors and doorways frequently appear in metaphorical or allegorical situations, literature and the arts, often as a portent of change.

### History

The earliest records are those represented in the paintings of the Egyptian tombs, in which they are shown as single or double doors, each in a single piece of wood. In Egypt, where the climate is intensely dry, there would be no fear of their warping, but in other countries it would be necessary to frame them, which according to Vitruvius (iv. 6.) was done with stiles (sea/si) and rails *(see: Frame and panel)*: the spaces enclosed being filled with panels (tympana) let into grooves made in the stiles and rails. The stiles were the vertical boards, one of which, tenoned or hinged, is known as the hanging stile, the other as the middle or meeting stile. The horizontal cross pieces are the top rail, bottom rail, and middle or intermediate rails. The most ancient doors were in timber, those made for King Solomon's temple being in olive wood (I Kings vi. 31-35), which were carved and overlaid with gold. The doors dwelt upon in Homer would appear to have been cased in silver or brass. Besides Olive wood, elm, cedar, oak and cypress were used. A 5,000-year-old door has been found by archaeologists in Switzerland. Many believe that one of the early purposes of doors was to keep the supernatural out, since ghosts can not go through doors.

All ancient doors were hung by pivots at the top and bottom of the hanging stile which worked in sockets in the lintel and sill, the latter being always in some hard stone such as basalt or granite. Those found at Nippur by Dr.Hilprecht, dating from 2000 B.C. were in dolerite. The tenons of the gates at Balawat were sheathed with bronze (now in the British Museum). These doors or gates were hung in two leaves, each about 8 ft 4 in (2.54 m) wide and 27 ft (8.2 m) high; they were encased with bronze bands or strips, 10 in. high, covered with repouss decoration of figures, etc. The wood doors would seem to have been about 3 in. thick, but the hanging stile was over 14 inches (360 mm) diameter. Other sheathings of various sizes in bronze have been found, which proves this to have been the universal method adopted to protect the wood pivots. In the Hauran in Syria, where timber is scarce the doors were made in stone, and one measuring 5 ft 4 in (1.63 m) by 2 ft 7 in (0.79 m) is in the British Museum; the band on the meeting stile shows that it was one of the leaves of a double door. At Kuffeir near Bostra in Syria, Burckhardt found stone doors, 9 to 10 ft (3.0 m). high, being the entrance doors of the town. In Etruria many stone doors are referred to by Dennis.

The ancient Greek and Roman doors were either single doors, double doors, sliding doors or folding doors, in the last case the leaves were hinged and folded back. In Eumachia, is a painting of a door with three leaves. In the tomb of Theron at Agrigentum there is a single four-panel door carved in stone. In the Blundell collection is a bas-relief of a temple with double doors, each leaf with five panels. Among existing examples, the bronze doors in the church of SS. Cosmas and Damiano, in Rome, are important examples of Roman metal work of the best period; they are in two leaves, each with two panels, and are framed in bronze. Those of the Pantheon are similar in design, with narrow horizontal panels in addition, at the top, bottom and middle. Two other bronze doors of the Roman period are in the Lateran Basilica.

The Greek scholar Heron of Alexandria created the earliest known automatic door in the 1st century AD during the era of Roman Egypt. The first foot-sensor-activated automatic door was made in China during the reign of Emperor Yang of Sui (r. 604–618), who had one installed for his royal library. The first automatic gate operators were later created in 1206 by the Arabic inventor, Al-Jazari.

Copper and its alloys were integral in medieval architecture. The doors of the church of the Nativity at Bethlehem (6th century) are covered with plates of bronze, cut out in patterns. Those of Hagia Sophia at Constantinople, of the 8th and 9th century, are wrought in bronze, and the west doors of the cathedral of Aix-la-Chapelle (9th century), of similar manufacture, were probably brought from Constantinople, as also some of those in St. Marks, Venice. The bronze doors on the Aachen Cathedral in Germany date back to about AD 800. Bronze baptistery doors at the Cathedral of Florence were completed in 1423 by Ghiberti.

Of the 11th and 12th centuries there are numerous examples of bronze doors, the earliest being one at Hildesheim, Germany (1015). The Hildesheim design affected the concept of Gniezno door in Poland.

Of others in South Italy and Sicily, the following are the finest: in Sant Andrea, Amalfi (1060); Salerno (1099); Canosa (1111); Troia, two doors (1119 and 1124); Ravello (1179), by Barisano of Trani, who also made doors for Trani cathedral; and in Monreale and Pisa cathedrals, by Bonano of Pisa. In all these cases the hanging stile had pivots at the top and bottom. The exact period when the hinge was substituted is not quite known, but the change apparently brought about another method of strengthening and decorating doors, viz, with wrought-iron bands of infinite varieties of design. As a rule three bands from which the ornamental work springs constitute the hinges, which have rings outside the hanging stiles fitting on to vertical tenons run into the masonry or wooden frame. There is an early example of the 12th century in Lincoln; in France the metal work of the doors of Notre Dame at Paris is perhaps the most beautiful in execution, but examples are endless throughout France and England.

Returning to Italy, the most celebrated doors are those of the Battistero di San Giovanni (Florence), which together with the door frames are all in bronze, the borders of the latter being perhaps the most remarkable: the modelling of the figures, birds and foliage of the south doorway, by Andrea Pisano (1330), and of the east doorway by Ghiberti (1425–1452), are of great beauty; in the north door (1402–1424) Ghiberti adopted the same scheme of design for the paneling and figure subjects in them as Andrea Pisano, but in the east door the rectangular panels are all filled, with bas-reliefs, in which Scripture subjects are illustrated with innumerable figures, these being probably the gates of Paradise of which Michelangelo speaks.

The doors of the mosques in Cairo were of two kinds; those which, externally, were cased with sheets of bronze or iron, cut out in decorative patterns, and incised or inlaid, with bosses in relief; and those in wood, which were framed with interlaced designs of the square and diamond, this latter description of work being Coptic in its origin. The doors of the palace at Palermo, which were made by Saracenic workmen for the Normans, are fine examples and in good preservation. A somewhat similar decorative class of door to these latter is found in Verona, where the edges of the stiles and rails are beveled and notched.

In the Renaissance period the Italian doors are quite simple, their architects trusting more to the doorways for effect; but in France and Germany the contrary is the case, the doors being elaborately carved, especially in the Louis XIV and Louis XV periods, and sometimes with architectural features such as columns and entablatures with pediment and niches, the doorway being in plain masonry. While in Italy the tendency was to give scale by increasing the number of panels, in France the contrary seems to have been the rule; and one of the great doors at Fontainebleau, which is in two leaves, is entirely carried out as if consisting of one great panel only.

The earliest Renaissance doors in France are those of the cathedral of St. Sauveur at Aix (1503). In the lower panels there are figures 3 ft (0.91 m). high in Gothicniches, and in the upper panels a double range of niches with figures about 2 ft (0.61 m). high with canopies over them, all carved in cedar. The south door of Beauvais Cathedral is in some respects the finest in France; the upper panels are carved in high relief with figure subjects and canopies over them. The doors of the church at Gisors (1575) are carved with figures in niches subdivided by classic pilasters superimposed. In St. Maclou at Rouen are three magnificently carved doors; those by Jean Goujon have figures in niches on each side, and others in a group of great beauty in the centre. The other doors, probably about forty to fifty years later, are enriched with bas-reliefs, landscapes, figures and elaborate interlaced borders.

When it comes to the world's largest door, there is not just one, in fact there are four and they all belong to NASA's Vehicle Assembly Building at the Kennedy Space Centre. The Vehicle Assembly Building was originally built to assembly of Apollo and Saturn vehicles and was then used to support Space Shuttle operations.

Each of the four doors are 139 metres or 456 feet high, in comparison the Statue of Liberty is only 93 metres or 305 feet high.

The oldest door in England can be found in Westminster Abbey and dates from 1050. In England in the 17th century the door panels were raised with bolection or projecting moldings, sometimes richly carved, round them; in the 18th century the moldings worked on the stiles and rails were carved with the egg and tongue ornament.

### Design and Styles

Many kinds of doors have specific names, depending on their purpose. The most common variety of door is the single-leaf door which consists of a single rigid panel that fills the doorway. Many variations on this basic design are possible, such as the double-leaf door or double doors and French windows that have two adjacent independent panels hinged on each side of the doorway. A half door or Dutch door or stable door is divided in half horizontally. Traditionally the top half can be opened to allow a horse or other animal to be fed, while the bottom half remained closed to keep the animal inside. This style of door has been adapted for homes.

Saloon doors are a pair of lightweight swing doors often found in public bars, and especially associated with the American west. Saloon doors, also known as cafe doors, often use bidirectional hinges which close the door regardless of which direction it is opened by incorporating springs. Saloon doors that only extend from knee-level to chest-level are known as batwing doors.

A blind door or Gibb door is a door with no visible trim or operable components. It is designed to blend with the adjacent wall in all finishes, and visually to be a part of the wall, a disguised door.

A barn door is a door characteristic of a barn. They are often/always found on barns, and because of a barn's immense size (often) doors are subsequently big for utility.

A French door is a door style consisting of a frame around one or more transparent and/or translucent panels (called lights or lites) that may be installed singly, in matching pairs, or even as series. A matching pair of these doors is called a French window as it resembles a door-height casement window. When a pair of French doors is used as a French window, the application does not generally include a central mullion (as do some casement window pairs), thus allowing a wider unobstructed opening. The frame typically requires a weather strip at floor level and where the doors meet to prevent water ingress. An espagnolette bolt may allow the head and foot of each door to be secured in one movement. The slender window joinery maximizes light into the room and minimizes the visual impact of the doorway joinery when considered externally. Often, the doors of a French window open outward onto a balcony, porch, or terrace and they may provide an entrance to a garden.

A louvred door has fixed or movable wooden fins (often called slats or louvers) which permit open ventilation while preserving privacy and preventing the passage of light to the interior. Being relatively weak structures, they are most commonly used for wardrobes and drying rooms, where security is of less importance than good ventilation, although a very similar structure is commonly used to form window shutters. Double louvred doors were introduced into *Seagate*, built in Florida in 1929 by Gwendolyn and Powel Crosley, that provided the desired circulation of air with an added degree of privacy in that it is impossible to see through the fins in any direction.

A composite door is a single leaf door that can be solid or with glass, and is usually filled with high density foam. Most composite doors carry *Secured by Design* accreditation and PAS 23 and PAS 24. A flush door is a completely smooth door, having plywood or MDF fixed over a light timber frame, the hollow parts of which are often filled with a cardboard core material. Skins can also be made out of hardboards, the first of which was invented by William H Mason in 1924. Called Masonite, its construction involved pressing and steaming wood chips into boards. Flush doors are most commonly employed in the interior of a dwelling, although slightly more substantial versions are occasionally used as exterior doors, especially within hotels and other buildings containing many independent dwellings.

A moulded door has the same structure as that of flush door. The only difference is that the surface material is a moulded skin made of MDF. Skins can also be made out of hardboards. A ledge and brace door is a door made from multiple vertical planks fixed together by two horizontal planks (the ledges) and kept square by a diagonal plank (the brace).

A wicket door is a pedestrian door built into a much larger door allowing access without requiring the opening of the larger door. Examples might be found on the ceremonial door of a cathedral or in a large vehicle door in a garage or hangar. A bifold door is a door unit that has several sections, folding in pairs. Wood is the most common material, and doors may also be metal or glass. Bifolds are most commonly made for closets, but may also be used as units between rooms.

A sliding glass door, sometimes called an Arcadia door or a Patio door, is a door made of glass that slides open and sometimes has a screen (a removable metal mesh that covers the door). Australian doors are a pair of plywood swinging doors often found in Australian public houses. These doors are generally red or brown in colour and bear a resemblance to the more formal doors found in other British Colonies' public houses. A false door is a wall decoration that looks like a door. In ancient Egyptian architecture, this was a common element in a tomb, the false door representing a gate to the afterlife. They can also be found in the funerary architecture of the desert tribes (e.g., Libyan Ghirza).

### Hinged Doors

Most doors are hinged along one side to allow the door to pivot away from the doorway in one direction but not in the other. The axis of rotation is usually vertical. In some cases, such as hinged garage doors, the axis may be horizontal, above the door opening.

Doors can be hinged so that the axis of rotation is not in the plane of the door to reduce the space required on the side to which the door opens. This requires a mechanism so that the axis of rotation is on the side other than that in which the door opens. This is sometimes the case in trains or airplanes, such as for the door to the toilet, which opens inward.

A swing doors has special hinges that allow it to open either outwards or inwards, and is usually sprung to keep it closed.

A selfbolting door is called as such because of its special hinges that permit the panel leaf to move laterally so that the door itself becomes a giant bolt for better security result. The selfbolting door principle can be used both for hinged doors as for rotating doors, as well as up-and-over doors (in the latter case, the bolts are then placed at top and bottom rather than at the sides).

French doors are derived from an original French design called the casement door. It is a door with lites where all or some panels would be in a casement door. A French door traditionally has a moulded panel at the bottom of the door. Called a French window when used in pairs as double-leaved doors with large glass panels in each door leaf, and in which the doors may swing out (most typical) as well as in.

Dutch door or stable door; the top half of the door operates independently from the bottom half. A variant exists in which opening the top part separately is possible, but because the lower part has a lip on the inside, closing the top part, while leaving the lower part open, is not. Garden door resembles a French window (with lites), but is more secure because only one door is operable. The hinge of the operating door is next to the adjacent fixed door and the latch is located at the wall opening jamb rather than between the two doors or with the use of an espagnolette bolt.

### Sliding Doors

It is often useful to have doors which slide along tracks, often for space or aesthetic considerations.

A bypass door is a door unit that has two or more sections. The doors can slide in either direction along one axis on parallel overhead tracks, sliding past each other. They are most commonly used in closets, in order to access one side of the closet at a time. The doors in a bypass unit will overlap slightly when viewed from the front, in order not to have a visible gap between them.

Doors which slide between two wall panels are called pocket doors. Sliding glass doors are common in many houses, particularly as an entrance to the backyard. Such doors are also popular for use for the entrances to commercial structures, although they are not counted as fire exit doors. The door that moves is called the "active leaf", while the door that remains fixed is called the "inactive leaf".

### Rotating Doors

A revolving door normally has several wings/leaves that hang on a central shaft forming compartments which rotate one way about a vertical axis. Doors using four wings are most common, but there are also examples with two, three, or six wings. The door may be motorised, or pushed manually using pushbars. People can walk out of and into the building at the same time. Between the point of access and the point of exit the user walks through an airlock. Revolving doors therefore create a good seal from the outside and help to reduce A/C and heating costs climate control from the building. This type of door is also often seen as a mark of prestige and glamour for a building and it not unusual for neighbouring buildings to install their own revolving doors when a rival building gets one.

A butterfly door is so called because of its two "wings". It consists of a double-wide panel with its rotation axle in the centre, effectively creating two separate openings when the door is opened. Butterfly doors are made to rotate open in one direction (usually counter clockwise), and rotate closed in the opposite direction. The door is not equipped with handles, so it is a "push" door. This is for safety, because if it could open in both directions, someone approaching the door might be caught off-guard by someone else opening the other side, thus impacting the first person. Such doors are popular in public transit stations, as it has a large capacity, and when the door is opened, traffic passing in both directions keeps the door open. They are particularly popular in underground subway stations, because they are heavy, and when air currents are created by the movement of trains, the force will be applied to both wings of the door, thus equalizing the force on either side, keeping the door shut.

### High Speed Door

A High Speed Door is a very fast door some with opening speeds of up to 4 m/s, mainly used in the industrial sector where the speed of a door has an effect on production logistics, temperature and pressure control. High Speed Clean Room Doors are used in Pharmaceutical industries for the special curtain and stainless steel frames. They guarantee the tightness of all accesses. The powerful high-speed doors have a smooth surface structure and no protruding edges.

Therefore, they can be easily cleaned and depositing of particles is largely excluded. High Speed Doors are made to handle a high number of openings, generally more than 200000 a year. They need to be built with heavy duty parts and counterbalance systems for speed enhancement and emergency opening function. The door curtain was originally made of PVC, but was later also developed in aluminium and acrylic glass sections. High Speed refrigeration and cold room doors with excellent insulation values was also introduced with the Green and Energy saving requirements.

In North America, the Door and Access Systems Manufacturing Association (DASMA) defines high-performance doors as non-residential, powered doors, characterized by rolling, folding, sliding or swinging action, that are either high-cycle (minimum 100 cycles/day) or high-speed (minimum 20 inches (508 mm)/second), and two out of three of the following: made-to-order for exact size and custom features, designed to be able to withstand equipment impact (break-away if accidentally hit by vehicle) or designed to sustain heavy usage with minimal maintenance.

### Others

Up-and-over or overhead doors are often used in garages. Instead of hinges it has a mechanism, often counterbalanced or sprung, that allows it to be lifted so that it rests horizontally above the opening. A roller shutter or sectional overhead door is one variant of this type.

A tambour door or roller door is an up-and-over door made of narrow horizontal slats and "rolls" up and down by sliding along vertical tracks and is typically found in entertainment centres and cabinets. Automatically opening doors are powered open and closed either by electricity, spring, or both. There are several methods by which an automatically opening door is activated:

1. A sensor detects traffic is approaching. Sensors for automatic doors are generally:
   - o A pressure sensor - e.g., a floor mat which reacts to the pressure of someone standing on it.

  - An infrared curtain or beam which shines invisible light onto sensors; if someone or something blocks the beam the door is triggered open.
  - A motion sensor which uses low-power microwave radar for the same effect.
  - A remote sensor (e.g. based on infrared or radio waves) can be triggered by a portable remote control, or is installed inside a vehicle. These are popular for garage doors.
2. A switch is operated manually, perhaps after security checks. This can be a push button switch or a swipe card.
3. The act of pushing or pulling the door triggers the open and close cycle. These are also known as power-assisted doors.

In addition to activation sensors automatically opening doors are generally fitted with safety sensors. These are usually an infrared curtain or beam, but can be a pressure mat fitted on the swing side of the door. The purpose of the safety sensor is to prevent the door from colliding with an object in its path by stopping or slowing its motion.

Inward opening doors are doors that can only be opened (or forced open) from outside a building. Such doors pose a substantial fire risk to occupants of occupied buildings when they are locked. As such doors can only be forced open from the outside, building occupants would be prevented from escaping. In commercial and retail situations manufacturers have included in the design a mechanism that allows an inward opening door to be pushed open outwards in the event of an emergency (which is often a regulatory requirement). This is known as a 'breakaway' feature. Pushing the door outward at its closed position, through a switch mechanism, disconnects power to the latch and allows the door to swing outward. Upon returning the door to the closed position, power is restored.

Rebated doors, a term chiefly used in Britain, are double doors having a lip or overlap (i.e. a Rabbet) on the vertical edge(s) where they meet. Fire-rating can be achieved with an applied edge-guard or astragal molding on the meeting stile, in accordance with the American Fire door.

### Applications

Architectural doors have numerous general and specialised uses. Doors are generally used to separate interior spaces (rooms, closets, etc.) for privacy, convenience, security, and safety reasons. Doors are also used to secure passages into a building from the exterior for reasons of safety and climate control. Doors also are applied in more specialised cases:

- A trapdoor is a door that is oriented horizontally in a floor or ceiling, often accessed via a ladder.
- Blast-proof doors are constructed to allow access to a structure but also to provide protection from the force of explosions.
- A garden door is any door that opens to a garden or backyard. This term is often used specifically for French windows, double French doors (with lites instead of panels), in place of a sliding glass door. The term also may refer to what is known as *patio doors*.
- A jib door is a concealed door, whose surface reflects the moldings and finishes of the wall. These were used in historic English houses, mainly as servants' doors.
- A pet door (also known as a dog door or cat flap) is an opening in a door to allow pets to enter and exit without the main door being opened. It may be simply covered by a rubber flap or it may be an actual door hinged on the top that the pet can push through. Pet doors may be mounted in a sliding glass door as a new (permanent or temporary) panel. Pet doors may be unidirectional, only allowing pets to exit. Pet doors may be electronic, only allowing pets with a special electronic tag to enter.

## Door Construction and Components

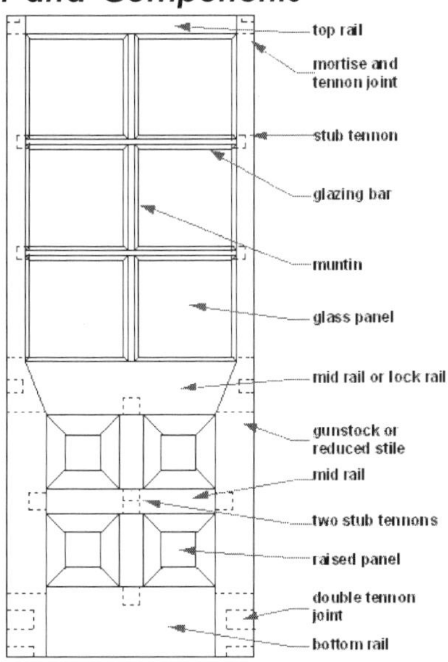


***Figure:*** *Parts of a panel and or glazed door*

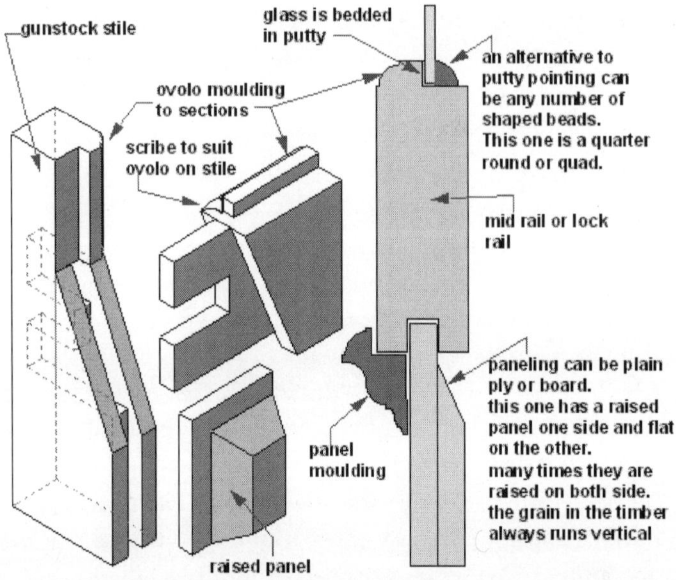


***Figure:*** *joint between midrail, lockrail and a gunstock stile*

***Figure:*** *A hollow door with one face removed*

## Panel doors

Panel doors, also called stile and rail doors, are built with frame and panel construction. EN 12519 is describing the terms which are officially used in European Member States. The main parts are listed below:

- Stiles - Vertical boards that run the full height of a door and compose its right and left edges. The hinges are mounted to the fixed side (known as the "hanging stile"), and the handle, lock, bolt, and/or latch are mounted on the swinging side (known as the "latch stile").
- Rails - Horizontal boards at the top, bottom, and optionally in the middle of a door that join the two stiles and split the door into two or more rows of panels.

  The "top rail" and "bottom rail" are named for their positions. The bottom rail is also known as "kick rail". A middle rail at the height of the bolt is known as the "lock rail", other middle rails are commonly known as "cross rails".
- Mullions - Smaller optional vertical boards that run between two rails, and split the door into two or more columns of panels, the term is used sometimes for verticals in doors, but more often (UK and Australia) it refers to verticals in windows.
- Muntin - Optional vertical members that divide the door into smaller panels.
- Panels - Large, wider boards used to fill the space between the stiles, rails, and mullions. The panels typically fit into grooves in the other pieces, and help to keep the door rigid. Panels may be flat, or in raised panel designs. Can be glued in or stay as a floating panel.
- Light or Lite - a piece of glass used in place of a panel, essentially giving the door a window.

### Plank and Batten Doors

Plank and batten doors are an older design consisting primarily of vertical slats:

- Planks - Vertical boards that extend the full height of the door, and are placed side by side filling the door's width.
- Battens - Smaller slats that extend horizontally across the door which the planks are affixed to.

  The battens hold the planks together. Sometimes a long diagonal slat or two are also implemented to prevent the door from skewing. On some doors, especially antique ones, the battens are replaced with iron bars that are often built into the hinges as extensions of the door-side plates.

### Ledged and Braced Doors

This type consists of vertical tongue and grooved boards held together with battens and diagonal braces.

### Impact-Resistant Doors

Impact-resistant doors have rounded stile edges to dissipate energy and minimize edge chipping, scratching and denting. The formed edges are often made of an engineered material such as Acrovyn. Impact-resistant doors excel in high traffic areas such as hospitals, schools, and hotels.

### Frame and Filled Doors

This type consists of a solid timber frame, filled on one face, face with Tongue and Grooved boards. Quite often used externally with the boards on the weather face.

### Flush Doors

Many modern doors, including most interior doors, are flush doors:

- Stiles and rails - As above, but usually smaller. They form the outside edges of the door.
- Core material: Material within the door used simply to fill space, provide rigidity and reduce druminess.
  - o Hollow-core - Often consists of a lattice or honeycomb made of corrugated cardboard, or thin wooden slats. Can also be built with staggered wooden blocks. Hollow-core flush doors are commonly used as interior doors.
    - – Lock block - A solid block of wood mounted within a hollow-core flush door near the bolt to provide a solid and stable location for mounting the door's hardware.
  - o Stave-core - Consists of wooden slats stacked upon one another in a manner similar to a plank & batten door (though the slats are usually thinner) or the wooden-block hollow-core (except that the space is entirely filled).
  - o Solid-core - Can consist of low-density particle board or foam used to completely fill the space within the door. Solid-core flush doors (especially foam-core ones) are commonly used as exterior doors because they provide more insulation and strength.
- Skin - The front and back faces of the door are then covered with wood veneer, thin plywood, sheet metal, fibreglass, or

vinyl. The wooden materials are usually layered with the grain alternating direction between layers to prevent warping. Fibreglass and metal-faced doors are sometimes given a layer of cellulose so that they may be stained to look like real wood.

### *Moulded Doors*

- Stiles and rails - As above, but usually smaller. They form the outside edges of the door.
- Core material: Material within the door used simply to fill space, provide rigidity and reduce druminess.
    - o Hollow-core - Often consists of a lattice or honeycomb made of corrugated cardboard, or thin wooden slats. Can also be built with staggered wooden blocks. Hollow-core flush doors are commonly used as interior doors.
        - – Lock block - A solid block of wood mounted within a hollow-core flush door near the bolt to provide a solid and stable location for mounting the door's hardware.
    - o Stave-core - Consists of wooden slats stacked upon one another in a manner similar to a plank & batten door (though the slats are usually thinner) or the wooden-block hollow-core (except that the space is entirely filled).
    - o Solid-core - Can consist of low-density particle board or foam used to completely fill the space within the door. Solid-core flush doors (especially foam-core ones) are commonly used as exterior doors because they provide more insulation and strength.
- Skin - The front and back faces of the door are covered with HDF / MDF skins.

### *Swing Direction*

Door swings For most of the world, door swings, or handing, are determined while standing on the outside or less secure side of the door while facing the door (i.e., standing on the side you use the key on, going from outside to inside, or from public to private).

It is especially important to get the hand and swing right on exterior doors, as the transom is usually sloped and sealed to resist water entry, and to properly drain. In some custom millwork (or with some master carpenters), the manufacture or installer will bevel the leading edge (the first edge to meet the jamb as the door is closing) so that the door fits tight without binding. Specifying an incorrect hand and/or swing will cause the door to bind, not close properly, or leak (for exterior doors). Fixing this specification error will be expensive and/or time consuming.

In North America, many doors now come with factory-installed hinges, pre-hung on the jamb and sills.

To determine hand, stand on the outside (or less secure) side of the door. While facing the door, if the hinge is on the right side of the door, the door is "Right handed"; or if the hinge is on the left, it is "Left handed".

If the door swings toward you, it is "Reverse swing"; or if the door swings away from you, it is "Normal swing".

In other words:

- Left hand hinge (LHH): Standing outside (or on the less secure side, or on the public side of the door), the hinges are on the left and the door opens in (away from you).
- Right hand hinge (RHH): Standing outside (or on the less secure side), the hinges are on the right and the door opens in (away from you).
- Left hand reverse (LHR): Standing outside the house (or on the less secure side), the hinges are on the left, knob on right, on opening the door it swings towards you (i.e. the door swings open towards the outside, or "outswing")
- Right hand reverse (RHR): Standing outside the house (i.e. on the less secure side), the hinges are on the right, knob on left, opening the door by pulling the door towards you (i.e. open swings to the outside, or "outswing")

***Note:*** In Australia, this is different. The refrigerator rule applies (you can't stand in a fridge, the door always opens towards you) - If the hinges are on the left then its a left hand (or left hung) door. If the hinges are on the right then its a right hand (or right hung) door.

In public buildings, exterior doors should open to the outside in order to comply with any fire codes that may be in force in that jurisdiction. If the door opens inward and there is a fire, there can be a crush of people who run for the door and they will not be able to open it.

### Dimensions

***United Kingdom:*** Single Internal Doors - ready made single leaf internal doors are often produced in imperial and metric sizes.

1. Range of UK imperial sizes for single internal doors (width x height):
   - o 15" x 78" = 1'3" x 6'6" = 381 x 1981 mm
   - o 18" x 78" = 1'6" x 6'6" = 457 x 1981 mm
   - o 21" x 78" = 1'9" x 6'6" = 533 x 1981 mm
   - o 24" x 78" = 2'0" x 6'6" = 610 x 1981 mm
   - o 27" x 78" = 2'3" x 6'6" = 686 x 1981 mm
   - o 30" x 78" = 2'6" x 6'6" = 762 x 1981 mm
   - o 32" x 80" = 2'8" x 6'8" = 813 x 2032 mm
   - o 33" x 78" = 2'9" x 6'6" = 838 x 1981 mm
   - o 34" x 78" = 2'10" x 6'6" = 864 x 1981 mm
   - o 36" x 78" = 3'0" x 6'6" = 914 x 1981 mm
2. Range of UK metric sizes for single leaf internal doors (width x height):
   - o 326 x 2040 mm
   - o 426 x 2040 mm
   - o 526 x 2040 mm
   - o 626 x 2040 mm
   - o 726 x 2040 mm
   - o 826 x 2040 mm
   - o 926 x 2040 mm

Double Internal Doors (French Doors) - pair of 2 single door leafs, usually rebated on the meeting stile.

- 36" x 78" = 3'0" x 6'6" = 915 x 1981 mm
- 42" x 78" = 3'6" x 6'6" = 1067 x 1981 mm
- 46" x 78" = 3'10" x 6'6" = 1168 x 1981 mm
- 48" x 78" = 4'0" x 6'6" = 1220 x 1981 mm
- 54" x 78" = 4'6" x 6'6" = 1372 x 1981 mm
- 60" x 78" = 5'0" x 6'6" = 1524 x 1981 mm

Single External Doors - ready made single leaf external doors.

- 30" x 78" = 2'6" x 6'6" = 761 x 1981 mm
- 32" x 80" = 2'8" x 6'8" = 813 x 2032 mm
- 33" x 78" = 2'9" x 6'6" = 838 x 1981 mm
- 34" x 82" = 2'10" x 6'10" = 864 x 2082 mm
- 36" x 84" = 3'0" x 7'0" = 915 x 2134 mm

### United States

A standard US residential door size is 36" x 80" (91 x 203 cm).

Exterior and passage (room to room) doors: Standard door sizes in the US are from 2'-6" to 3'-0" wide, increasing in 2" increments. Most residential interior doors are 2'-6" (0.76 m) wide except when designed to allow wheelchair access, then 3'-0" (0.91). Residential doors are often 6'-8" high, as are many small stores, offices, and other light commercial buildings. Larger commercial, public buildings and grand homes often use doors of greater height. Older buildings often have smaller doors.

***Closets:*** small spaces such as closets, dressing rooms, half-baths, storage rooms, cellars, etc. often are accessed through doors smaller than passage doors in one or both dimensions but similar in design.

***Garages:*** Garage doors are generally 7'-0" or 8'-0" wide for a single-car opening.

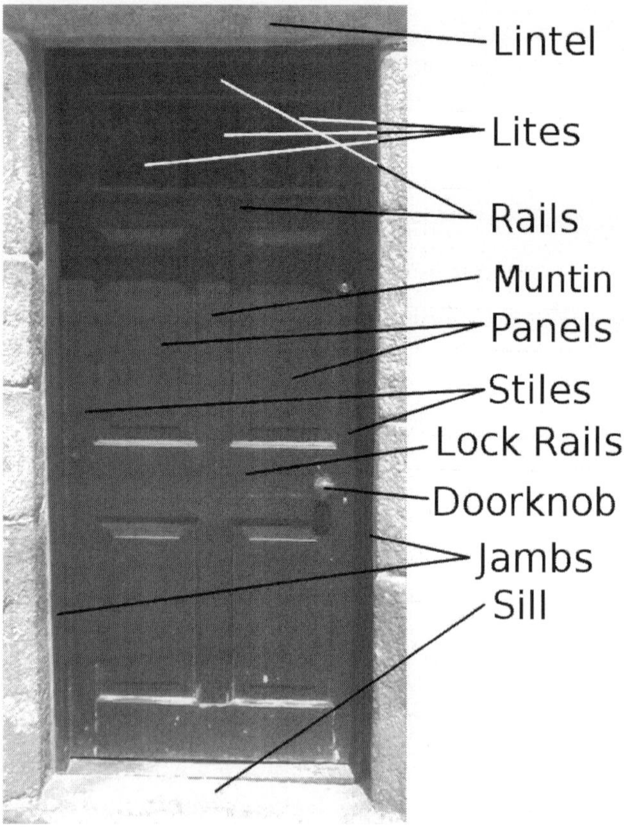


***Figure:*** *A diagram illustrating the components of a panel door*

## Doorway Components

When framed in wood for snug fitting of a door, the doorway consists of two vertical *jambs* on either side, a *lintel* or *head jamb* at the top, and perhaps a *threshold* at the bottom.

When a door has more than one movable section, one of the sections may be called a *leaf*.

- Lintel - A horizontal beam above a door that supports the wall above it. (Also known as a *header*)
- Jambs - The vertical posts that form the sides of a door frame, where the hinges are mounted, and with which the bolt interacts.
- Sill - A horizontal beam below the door that supports the frame
- Doorstop - a thin slat built inside the frame to prevent a door from swinging through when closed, which might break the hinges.
- Architrave - The decorative molding that outlines a door frame. (called an Archivolt if the door is arched). Called *door casing* or *brickmold* in North America.
- Door knocker is an ornamental object hinged to the outside surface of a door and operated (by giving a knock on the door) to get the attention of residents in the room to gain admittance. As a verb, a door knock is a tapping action performed upon the outside of a door in order to signal to those within that one wishes to gain entry. One stylized form of knock is the call of "Shave and a Haircut".

## Related Hardware

Door furniture or hardware refers to any of the items that are attached to a door or a drawer to enhance its functionality or appearance. This includes items such as hinges, handles, door stops, etc.

## CE Marking

Doors applied for external usage with no fire and/or smoke resisting characteristics are CE marked by hEN 14351-1. Soon prEN 16034 will be publicly available aiming to cover Fire and Smoke characteristics as well as some additional ones which relate to these two. prEN 14351-2 is covering internal pedestrian doors.

Manufacturer may be assisted by going through the short guideline which was developed by both System House and Manufacturer associations.

### *Door-Related Accidents*

Door safety relates to prevention of door-related accidents. Such accidents take place in various forms, and in a number of locations; ranging from car doors to garage doors.

Contrast *door safety* with *door security*.

### *Severity and Frequency of Accidents*

Accidents vary in severity and frequency. According to the National Safety Council in the United States, 300,000 injuries are caused by doors every year.

The types of accidents vary from relatively minor cases where doors cause damage to other objects, such as walls, to serious cases resulting in human injury, particularly to fingers. A closing door can exert up to 40 tons per square inch of pressure between the hinges. Because of the number of accidents taking place, there has been a surge in the number of lawsuits. Thus organisations may be at risk when car doors or doors within buildings are unprotected.

***According to the US General Services Administration:*** "...It is essential that children's fingers be protected from being crushed or otherwise injured in the hinge space of a swinging door or gate. There are simple devices available to attach to the hinge side, ensuring that this type of injury does not occur. As the door closes, the hand is pushed out of the opening, away from harm.

In addition, young children are vulnerable to injury when they fall against the other (hinged) side of doors and gates, striking projected hinges. Piano hinges are not recommended to alleviate this problem as they tend to sag over time with heavy use. Instead, an inexpensive device fitting over hinges is available on the market and should be used to ensure safety..."

### *Accident Types and Prevention*

***Outward and Inward Opening:*** Whenever a door is opened outwards there is a risk that it could strike another person. In many cases this can be avoided by architectural design which favours doors which open inwards into rooms (from the perspective of a common area such as a corridor, the door opens outwards). In cases where this is infeasible, it may be possible to avoid an accident by placing windows in the door.

However, inward-hinged doors can also escalate an accident by preventing people from escaping the building: people inside the building may press against the doors, and thus prevent the doors from opening. This was partly the case in the Grue Church fire in Norway in 1822.

Today, the exterior doors of most large (especially public) buildings open outward, while interior doors such as doors to individual rooms, offices, suites, etc. open inward, as do many exterior doors of houses, particularly in North America.

### Doorstops

Doorstops are simple devices used to prevent a door from coming into contact with another object (typically a wall). Without the door stop damage might be done to the wall. They may either absorb the force of a moving door, or hold the door in place to prevent unintended motion.

### Door Guards

The purpose of door guards (also known as hinge guards, anti-finger trapping devices, or finger guards) is to reduce the number of finger trapping accidents in doors, as doors pose a risk to children especially when closing.

Door guards protect fingers in door hinges by covering the gap that is created by opening doors by covering the hinges of doors with a piece of rubber or plastic that wraps from the door frame to the door. There are also door safety products which eject the fingers from the push side of the door as it is being closed.

There are various levels of door protection. Front door protection a front anti-finger trapping device but leaves the rear hinge pin side of the door unprotected. Full door protection uses front and rear anti-finger trapping devices and ensures the hinge side of a door is fully protected. Which level of protection is appropriate should be determined by a risk assessment of the door.

There is also handle-side door protection, which prevents the door from slamming shut on the frame, which can cause injury to fingers/hands.

### Safety Doors

A safety door prevents finger injuries near the hinges without the use of door guards. Rather than cover the danger area, the approach is to change the shape of the door so that an accessible gap does not form in the first place. This is achieved by adding a perfectly circular ("bull-nose" shaped) extension to the door, which moves in and out of a cavity as the door opens and closes. This prevents any part of a hand being crushed near the hinges - either inside or outside. These doors have an operating range of slightly over 90 degrees, so their use is limited to where they come into contact with a side wall when fully open (or where they can be prevented from opening too far by a doorstop).

### Glass Doors

Glass doors pose the risk of unintentional collision if a person believes the door to be open when it is closed, or is unaware there is a door at all. This risk may be particularly pronounced with sliding glass doors because they often have large single panes which are hard to see. To prevent injury from glass doors, stickers or other types of warnings are sometimes placed on the glass surface to make it more visible.

For instance, in the UK, Regulation 14 of the Workplace (Health and Safety Regulations) 1992 requires the marking of windows and glass doors to make them conspicuous. Australian Standards: AS1288 and AS2208 require glass doors to be made from laminated or toughened glass.

### Fire

Special purpose fire doors are often employed in buildings to reduce the overall risk of fire, particularly by preventing the spread of fire and smoke. In cases where they are improperly installed, employed, or tampered with, the risk of fire can be increased. Door closers are sometimes used to ensure fire doors remain closed.

An additional risk in a fire is that doors may prevent access to emergency services personnel in order to fight the fire, rescue occupants, etc. Door breaching techniques may be required in these situations to gain access. Panic bars are often used in buildings so that a door locked from the exterior can quickly and easily be opened from the inside in the event of a fire or other emergency.

### Vehicle Doors

There may be an increased risk of trapping hands or fingers in car doors compared to other types of doors. In some car accidents, injury to occupants from the movement of car doors may occur.

Bicyclists often fear collision with an opening car door in case the car's occupant does not look carefully to check that it is safe to open the door. Because cyclists often ride near parked cars along the side of the road they are particularly vulnerable to this.

### Aircraft Doors

Doors which lead from interior, pressurized, sections of an aircraft to exterior or unpressurized areas can pose extreme risk if they are inadvertently opened during flight. This can be mitigated by having doors that open inwardly and are designed to be forced into their door frames by the internal cabin pressure - most cabin doors are of this type. However, an outward opening door is often advantageous for cargo doors to maximise available space, and these need to be secured by hefty locking mechanisms to overcome internal pressure.

A number of accidents have occurred where outward-opening aircraft doors were opened in flight, often accidentally:

- American Airlines Flight 96 (1972)
- Turkish Airlines Flight 981 (1974)
- Pan Am Flight 125 (1987)
- United Airlines Flight 811 (1989)

### Window

***Figure:*** *Window of traditional design in Porto Covo, Portugal*

A window is a transparent or translucent opening in a wall, door or vehicle that allows the passage of light and, if not closed or sealed, air and sound. Windows are usually glazed or covered in some other transparent or translucent material like float glass. Windows are held in place by frames. Many glazed windows may be opened, to allow ventilation, or closed, to exclude inclement weather.

### Etymology

The word *window* originates from the Old Norse 'vindauga', from 'vindr – wind' and 'auga–eye', i.e., *wind eye.* In Norwegian Nynorsk and Icelandic the Old Norse form has survived to this day (in Icelandic only as a less used synonym to *gluggi*), in Swedish the word *vindöga* remains as a term for a hole through the roof of a hut, and in the Danish language 'vindue' and Norwegian Bokmål 'vindu', the direct link to 'eye' is lost, just like for 'window'. The Danish (but not the Bokmål) word is pronounced fairly similarly to *window.*

*Window* is first recorded in the early 13th century, and originally referred to an unglazed hole in a roof. *Window* replaced the Old English *eagþyrl,* which literally means 'eye-hole,' and 'eagduru' 'eye-door'. Many Germanic languages however adopted the Latin word 'fenestra' to describe a window with glass, such as standard Swedish 'fönster', or German 'Fenster'. The use of *window* in English is probably because of the Scandinavian influence on the English language by means of loanwords during the Viking Age. In English the word *fenester* was used as a parallel until the mid-18th century and *fenestration* is still used to describe the arrangement of windows within a façade. Also, words such as "defenestration" are in use, meaning to throw something out of a window.

***From Webster's 1828 Dictionary:*** *Window,* n. [G. The vulgar pronunciation is windor, as if from the Welsh gwyntdor, wind-door.]

### History

The earliest windows were just holes in a wall. Later, windows were covered with animal hide, cloth, or wood. Shutters that could be opened and closed came next. Over time, windows were built that both protected the inhabitants from the elements and transmitted light: mullioned glass windows, which joined multiple small pieces of glass with leading, paper windows, flattened pieces of translucent animal horn, and plates of thinly sliced marble. In the Far East, paper was used to fill windows.

The Romans were the first known to use glass for windows, a technology likely first produced in Roman Egypt—In Alexandria ca. 100 AD, cast glass windows, albeit with poor optical properties, began to appear—but these were small thick productions, little more than blown glass jars (cylindrical shapes) flattened out into sheets with circular striation patterns throughout. It would be over a millennium before a window glass became transparent enough to see through clearly, as we think of it now.

***Figure:*** *Alabaster 'mullion' divided decorative windows in Santa Maria La Major church (Morella, Spain).*

Over the centuries techniques were developed to shear through one side of a blown glass cylinder and produce thinner rectangular window panes from the same amount of glass material. This gave rise to tall narrow windows, usually separated by a vertical support called a mullion. Mullioned glass windows were the windows of choice among European well-to-do, whereas paper windows were economical and widely used in ancient China, Korea and Japan.

In England, glass became common in the windows of ordinary homes only in the early 17th century whereas windows made up of panes of flattened animal horn were used as early as the 14th century. Noted science historian, author and television show host/producer James Burke attributes the rapid deforestation of Great Britain in the late 1500s to the uptick in production of glazed windows as well as iron cannon production (1st Cast in 1547). He writes further this gave rise to coal for fuel, which spurred iron production, requiring more coal, and more iron, then steam engine pumps, canals... and more iron; all because windows became a middle class commodity in the latter days of the little ice age, one large factor among several leading to the deforesting English woodlands, and the switch over to a coal economy.

Modern-style floor-to-ceiling windows became possible only after the industrial plateglass making processes were perfected. Modern windows are usually filled with glass, although a few are transparent plastic.

### Types

***Fixed:*** A window that cannot be opened, whose function is limited to allowing light to enter (Unlike an unfixed window, which can open and close). Clerestory windows are often fixed. Transom windows may be fixed or operable. This type of window is used in situations where light or vision alone is needed as no ventilation is possible windows without the use of trickle vents or overglass vents.

### Single-Hung Sash

One sash is movable (usually the bottom one) and the other fixed. This is the earlier form of sliding sash window, and is also cheaper.

### Double-Hung Sash

A sash window is the traditional style of window in the United Kingdom, and many other places that were formerly colonized by the UK, with two parts (sashes) that overlap slightly and slide up and down inside the frame. The two parts are not necessarily the same size. Currently most new double-hung sash windows use spring balances to support the sashes, but traditionally, counterweights held in boxes on either side of the window were used. These were and are attached to the sashes using pulleys of either braided cord or, later, purpose-made chain. Three types of spring balances are called a tape or clock spring balance; channel or block-and-tackle balance; and a spiral or tube balance.

Double-hung sash windows were traditionally often fitted with shutters. Sash windows can be fitted with simplex hinges that let the window be locked into hinges on one side, while the rope on the other side is detached—so the window can be opened for fire escape or cleaning.

### Horizontal Sliding Sash

Has two or more sashes that overlap slightly but slide horizontally within the frame. In the UK, these are sometimes called *Yorkshire* sash windows, presumably because of their traditional use in that county.

### Casement

***Figure:*** *Casement window*

A window with a hinged sash that swings in or out like a door comprising either a side-hung, top-hung (also called "awning window"), or occasionally bottom-hung sash or a combination of these types, sometimes with fixed panels on one or more sides of the sash. In the USA, these are usually opened using a crank, but in parts of Europe they tend to use projection friction stays and espagnolette locking. Formerly, plain hinges were used with a casement stay. Handing applies to casement windows to determine direction of swing; a casement window may be left-handed, right-handed, or double. The casement window is the dominant type now found in the UK and parts of Europe.

### Awning

An awning window is a casement window that is hung horizontally, hinged on top, so that it swings outward like an awning. Emilie Poisson designed this window.

### Hopper

A hopper window is a bottom-pivoting casement window that opens by tilting vertically, typically to the inside. (Mostly used for schools)

### Tilt and Slide

A window (more usually a door-sized window) where the sash tilts inwards at the top and then slides horizontally behind the fixed pane.

### Tilt and Turn

A tilt and turn *window can both tilts inwards at the top or opens inwards from hinges at the side. This is the most common type of window in Germany, its country of origin. It is also widespread in many other European countries.*

### Transom

A window above a door; in an exterior door the transom window is often fixed, in an interior door it can open either by hinges at top or bottom, or rotate on hinges. It provided ventilation before forced air heating and cooling. A fan-shaped transom is known as a fanlight, especially in the British Isles.

### Jalousie Window

Also known as a louvered window, the jalousie window consists of parallel slats of glass or acrylic that open and close like a Venetian blind, usually using a crank or a lever. They are used extensively in tropical architecture. A jalousie door is a door with a jalousie window.

### Clerestory

A window set in a roof structure or high in a wall, used for daylighting.

### Skylight

A flat or slope window used for daylighting, built into a roof structure that is out of reach.

### Roof

A sloped window used for daylighting, built into a roof structure. It is one of the few windows that could be used as an exit. Larger roof windows meet building codes for emergency evacuation.

### Roof Lantern

A roof lantern is a multi-paned glass structure, resembling a small building, built on a roof for day or moon light. Sometimes includes an additional clerestory. May also be called a cupola.

***Figure:*** *Bay window*

### Bay

A multi-panel window, with at least three panels set at different angles to create a protrusion from the wall line.

### Oriel

This form of bay window most often appears in Tudor-style houses and monasteries.

It projects from the wall and does not extend to the ground. Originally a form of porch, they are often supported by brackets or corbels.

### Thermal

Thermal, or Diocletian, windows are large semicircular windows (or niches) which are usually divided into three lights (window compartments) by two mullions. The central compartment is often wider than the two side lights on either side of it.

### Picture

A picture window is a large fixed window in a wall, typically without glazing bars, or glazed with only perfunctory glazing bars near the edge of the window. Picture windows provide an unimpeded view, as if framing a picture.

### *Multi-Lit*

A window glazed with small panes of glass separated by wooden or lead *glazing bars;;, or "muntins*, arranged in a decorative *glazing pattern* often dictated by the building's architectural style. Due to the historic unavailability of large panes of glass, the multi-lit (or *lattice window*) was the most common window style until the beginning of the 20th century, and is still used in traditional architecture.

### *Emergency Exit/Egress*

A window big enough and low enough so that occupants can escape through the opening in an emergency, such as a fire. In many countries, exact specifications for emergency windows in bedrooms are given in many building codes. Specifications for such windows may also allow for the entrance of emergency rescuers. Vehicles, such as buses and aircraft, frequently have emergency exit windows as well.

### *Stained Glass*

A window composed of pieces of coloured glass, transparent, translucent or opaque, frequently portraying persons or scenes. Typically the glass in these windows is separated by lead glazing bars. Stained glass windows were popular in Victorian houses and some Wrightian houses, and are especially common in churches.

### *French*

A French window (when hinged *French door*) is a large door-sized lattice light, typically set in pairs or multiples thereof. Known as *porte-fenêtre* in France and *portafinestra* in Italy, they often overlook a terrace. Therefore, these windows are use in modern houses.

### *Double Paned*

Double paned windows have two parallel panes (slabs of glass) with a separation of typically about 1 cm; this space is permanently sealed and filled at the time of manufacture with dry air or other dry nonreactive gas. Such windows provide a marked improvement in thermal insulation (and usually in acoustic insulation as well) and are resistant to fogging and frosting caused by temperature differential. They are widely used for residential and commercial construction in intemperate climates.

Triple-paned windows have been commercially manufactured and marketed with claims of additional benefit but have not become common.

## Terms

EN 12519 is the European norm that describes windows terms officially used in EU Member States. The main terms are:

***Figure:*** *Casement window, with latticed lights*

- Light is the area between the outer parts of a window (transom, sill and jambs), usually filled with a glass pane. Multiple panes are divided by mullions when load-bearing, muntins when not.
- Lattice light is a compound window pane madeup of small pieces of glass held together in a lattice.
- Fixed window is a unit of one non-moving lite. The terms *single-light, double-light,* etc., refer to the number of these glass panes in a window.
- Sash unit is a window consisting of at least one sliding glass component, typically composed of two lites (known as a *double-light*).
- Replacement window in the United States means a framed window designed to slip inside the original window frame from the inside after the old sashes are removed. In Europe it usually means a complete window including a replacement outer frame.
- New construction window, in the US, means a window with a nailing fin that is inserted into a rough opening from the outside before applying siding and inside trim. A nailing fin is a projection on the outer frame of the window in the same plane as the glazing, which overlaps the prepared opening, and can thus be 'nailed' into place.

In the UK and mainland Europe, windows in new-build houses are usually fixed with long screws into expanding plastic plugs in the brickwork.

A gap of up to 13mm is left around all four sides, and filled with expanding polyurethane foam.

This makes the window fixing weatherproof but allows for expansion due to heat.

- Lintel is a beam over the top of a window, also known as a transom.
- window-sill is the bottom piece in a window frame. Window sills slant outward to drain water away from the inside of the building.
- Secondary Glazing is an additional frame applied to the inside of an existing frame, usually used on protected or listed buildings to achieve higher levels of thermal and sound insulation without compromising the look of the building
- Decorative millwork is the moulding, cornices and lintels often decorating the surrounding edges of the window.

### *Labelling*

The United States NFRC Window Label lists the following terms:

- Thermal transmittance (U-factor), best values are around U-0.15 (equal to 0.8 W/m2/K)
- Solar heat gain coefficient (SHGC), ratio of solar heat (infrared) passing through the glass to incident solar heat
- Visible transmittance (VT), ratio of transmitted visible light divided by incident visible light
- Air leakage (AL), measured in cubic foot per minute per linear foot of crack between sash and frame
- Condensation resistance (CR), measured between 1 and 100 (the higher the number, the higher the resistance of the formation of condensation)

The European harmonised standard hEN 14351-1, which deals with doors and windows, defines 23 characteristics (divided into *essential* and *non* essential.

*Two other, preliminary European Norms that are under development deal with internal pedestrian doors (prEN 14351-2), smoke and fire resisting doors, and openable windows (prEN 16034).*

## Window Construction

***Figure:*** *5-chamber plastic window profile*

Windows can be a significant source of heat transfer. Therefore, insulated glazing units consist of two or more panes to reduce the transfer of heat.

## Grids or Muntins

These are the pieces of framing that separate a larger window into smaller panes. In older windows, large panes of glass were quite expensive, so muntins let smaller panes fill a larger space.

In modern windows, light-coloured muntins still provide a useful function by reflecting some of the light going through the window, making the window itself a source of diffuse light (instead of just the surfaces and objects illuminated within the room).

By increasing the indirect illumination of surfaces near the window, muntins tend to brighten the area immediately around a window and reduce the contrast of shadows within the room.

### Frame and Sash Construction

| *Material* | *thermal resistance* | *Durability* | *Maintenance* | *Cost* | *Recycled content* | *Comment* |
|---|---|---|---|---|---|---|
| Wood | very good | variable | high | high | low | shrinks and swells with humidity changes |
| PVC ("vinyl") | very good | good* | low | low | very low | |
| Aluminum | bad** | good | very low | low | typically > 95% | used in most large structures |
| Steel | medium | superior | very low | high | > 98% | typically welded at corner joints |
| Fiberglass | very good | very good* | very low | high | medium | |

* PVC and fibreglass frames perform well in accelerated weathering tests. Because PVC is not as strong as other materials, some PVC frames are reinforced with metal or composite materials to improve their structural strength.

** Modern aluminium window frames are typically separated by a thermal break made of a polyamide. This greatly increases thermal resistance, while retaining virtually all of the structural strength.

Composites may combine materials to obtain aesthetics of one material with the functional benefits of another.

***Figure:*** *A typical installation of insulated glazing windows with uPVC window frames.*

A special class of PVC window frames, uPVC window frames, became widespread since the late 20th century, particularly in Europe: there were 83.5 million installed by 1998 with numbers still growing as of 2012.

## Glazing and Filling

Low-emissivity coated panes reduce heat transfer by radiation, which, depending on which surface is coated, helps prevent heat loss (in cold climates) or heat gains (in warm climates). High thermal resistance can be obtained by evacuating or filling the insulated glazing units with gases such as argon or krypton, which reduces conductive heat transfer due to their low thermal conductivity. Performance of such units depends on good window seals and meticulous frame construction to prevent entry of air and loss of efficiency.

Modern double-pane and triple-pane windows often include one or more low-e coatings to reduce the window's U-factor (its insulation value, specifically its rate of heat loss). In general, soft-coat low-e coatings tend to result in a lower solar heat gain coefficient (SHGC) than hard-coat low-e coatings.

Modern windows are usually glazed with one large sheet of glass per sash, while windows in the past were glazed with multiple panes separated by *glazing bars*, or *muntins*, due to the unavailability of large sheets of glass. Today, glazing bars tend to be decorative, separating windows into small panes of glass even though larger panes of glass are available, generally in a pattern dictated by the architectural style at use. Glazing bars are typically wooden, but occasionally lead glazing bars soldered in place are used for more intricate glazing patterns.

## Other Construction Details

Many windows have movable window coverings such as blinds or curtains to keep out light, provide additional insulation, or ensure privacy. Windows allow natural light to enter, but too much can have negative effects such as glare and heat gain. Additionally, while windows let the user see outside, there must be a way to maintain privacy on in the inside. Window coverings are practical accommodations for these issues.

## Windows and the Sun

***Sun Incidence Angle:*** Historically, windows are designed with surfaces parallel to vertical building walls. Such a design allows considerable solar light and heat penetration due to the most commonly occurring incidence of sun angles. In passive solar building design, an extended eave is typically used to control the amount of solar light and heat entering the window(s).

An alternate method is to calculate an optimum window mounting angle that accounts for summer sun load minimization, with consideration of actual latitude of the building. This process has been implemented, for example, in the Dakin Building in Brisbane, California—in which most of the fenestration is designed to reflect summer heat load and help prevent summer interior over-illumination and glare, by canting windows to nearly a 45 degree angle.

### Solar Window

Photovoltaïc windows not only provide a clear view and illuminate rooms, but also convert sunlight to electricity for the building. In most cases, translucent photovoltaïc cells are used.

### Passive Solar

Passive solar windows allow light and solar energy into a building while minimizing air leakage and heat loss. Properly positioning these windows in relation to sun, wind, and landscape—while properly shading them to limit excess heat gain in summer and shoulder seasons, and providing thermal mass to absorb energy during the day and release it when temperatures cool at night—increases comfort and energy efficiency. Properly designed in climates with adequate solar gain, these can even be a building's primary heating system.

### Window Coverings

A window covering is a shade or screen that provides multiple functions. For example, some window coverings control solar heat gain and glare. There are external shading devices and internal shading devices. Low-e window film is a low-cost alternative to window replacement to transform existing poorly-insulating windows into energy-efficient windows. For high-rise buildings, smart glass can provide an alternative.

## Brick

A brick is a block or a single unit of a ceramic material used in masonry construction. Typically bricks are stacked together or laid as brickwork using various kinds of mortar to hold the bricks together and make a permanent structure. Bricks are typically produced in common or standard sizes in bulk quantities. They have been regarded as one of the longest lasting and strongest building materials used throughout history.

In the general sense, a "brick" is a standard-sized weight-bearing building unit. Bricks are laid in horizontal courses, sometimes dry and sometimes with mortar. When the term is used in this sense, the brick might be made from clay, lime-and-sand, concrete, or shaped stone.

In a less clinical and more colloquial sense, bricks are made from dried earth, usually from clay-bearing subsoil. In some cases, such as adobe, the brick is merely dried. More commonly it is fired in a kiln of some sort to form a true ceramic.

### History

***Middle East:*** The earliest bricks were *dried brick*, meaning they were formed from clay-bearing earth or mud and dried (usually in the sun) until they were strong enough for use. The oldest discovered bricks, originally made from shaped mud and dating before 7500 BC, were found at Tell Aswad, in the upper Tigris region and in southeast Anatolia close to Diyarbakir. Other more recent findings, dated between 7,000 and 6,395 BC, come from Jericho, CatalHüyük, and the ancient Indus Valley cities of Buhen, Mohenjo-daro, Harappa, and Mehrgarh.

Ceramic, or *fired brick* was used as early as 4500 BC in early Indus Valley cities.

### Rome

The Romans made use of fired bricks, and the Roman legions, which operated mobile kilns, introduced bricks to many parts of the empire. Roman bricks are often stamped with the mark of the legion that supervised their production. The use of bricks in southern and western Germany, for example, can be traced back to traditions already described by the Roman architect Vitruvius.

### Europe

The oldest domestic bricks were found in Greece. In the 12th century, bricks from Northern-Western Italy were re-introduced to Northern Germany, where an independent tradition evolved.

It culminated in the so-called brick Gothic, a reduced style of Gothic architecture that flourished in Northern Europe, especially in the regions around the Baltic Sea, which are without natural rock resources.

Brick Gothic buildings, which are built almost exclusively of bricks, are to be found in Denmark, Germany, Poland, and Russia.

During the Renaissance and the Baroque, visible brick walls were unpopular and the brickwork was often covered with plaster. It was only during the mid-18th century that visible brick walls regained some degree of popularity, as illustrated by the Dutch Quarter of Potsdam, for example.

The transport in bulk of building materials such as bricks over long distances was rare before the age of canals, railways, roads and heavy goods vehicles. Before this time bricks were generally made close to their point of intended use. It has been estimated that in England in the 18th century carrying bricks by horse and cart for ten miles (approx. 16 km) over the poor roads then existing could more than double their price.

Bricks were often used for reasons of speed and economy, even in areas where stone was available. The buildings of the Industrial Revolution in Britain were largely constructed of brick and timber due to the demand created. During the building boom of the 19th century in the eastern seaboard cities of Boston and New York City, for example, locally made bricks were often used in construction in preference to the brownstones of New Jersey and Connecticut for these reasons.

The trend of building high office buildings that emerged towards the beginning of the 20th century displaced brick in favour of cast and wrought iron and later steel and concrete. Some early 'skyscrapers' were made in masonry, and demonstrated the limitations of the material – for example, the Monadnock Building in Chicago (opened in 1896) is masonry and just 17 stories high; the ground walls are almost 6 feet (1.8 m) thick to give the needed support; clearly building any higher would lead to excessive loss of internal floor space on the lower floors. Brick was revived for high structures in the 1950s following work by the Swiss Federal Institute of Technology and the Building Research Establishment in Watford, UK.

This method produced 18-story structures with load-bearing walls no thicker than a single brick (150–225 mm). This potential has not been fully developed because of the ease and speed in building with other materials; in the late 20th century brick was confined to low- or medium-rise structures or as a thin decorative cladding over concrete-and-steel buildings or for internal non-load-bearing walls.

In Victorian London, bright red brick was chosen to make buildings more visible in the heavy fog that caused transport problems.

## China

In pre-modern China, brick-making was the job of a lowly and unskilled artisan, but a kiln master was respected as a step above the former. Early traces of bricks were found in a ruin site in Xi'an in 2009 dated back about 3800 years ago. Before this discovery, it is widely believed that bricks appeared about 3000 years ago in the Western Zhou dynasty since the earliest bricks were found in Western Zhou ruins. These bricks are the earliest bricks discovered that were made by a fired process. Early descriptions of the production process and glazing techniques used for bricks can be found in the Song Dynasty carpenter's manual *YingzaoFashi*, published in 1103 by the government official Li Jie, who was put in charge of overseeing public works for the central government's construction agency. The historian Timothy Brook writes of the production process in Ming Dynasty China (aided with visual illustrations from the *TiangongKaiwu* encyclopedic text published in 1637):

***Figure:*** *The brickwork of Shebeli Tower in Iran displays 12th-century craftsmanship*

> *...the kilnmaster had to make sure that the temperature inside the kiln stayed at a level that caused the clay to shimmer with the colour of molten gold or silver. He also had to know when to quench the kiln with water so as to produce the surface glaze. To anonymous labourers fell the less skilled stages of brick production: mixing clay and water, driving oxen over the mixture to trample it into a thick paste, scooping the paste into standardized wooden frames (to produce a brick roughly 42 cm long, 20 cm wide, and 10 cm thick), smoothing the surfaces with a wire-strung bow, removing them from the frames, printing the fronts and backs with stamps that indicated where the bricks came from and who made them, loading the kilns with fuel (likelier wood than coal), stacking the bricks in the kiln, removing them to cool while the kilns were still hot, and bundling them into pallets for transportation. It was hot, filthy work.*

The idea of signing the worker's name and birth date on the brick and the place where it was made was not new to the Ming era and had little or nothing to do with vanity. As far back as the Qin Dynasty (221 BC–206 BC), the government required blacksmiths and weapon-makers to engrave their names onto weapons in order to trace the weapons back to them, lest their weapons should prove to be of a lower quality than the standard required by the government.

### Methods of Manufacture

Modern clay bricks are formed in one of three processes – soft mud, dry press, or extruded.

Normally, brick contains the following ingredients:

1. Silica (sand) – 50% to 60% by weight
2. Alumina (clay) – 20% to 30% by weight
3. Lime – 2 to 5% by weight
4. Iron oxide – ≤ 7% by weight
5. Magnesia – less than 1% by weight

### Mud Bricks

The soft mud method is the most common, as it is the most economical. It starts with the raw clay, preferably in a mix with 25–30% sand to reduce shrinkage. The clay is first ground and mixed with water to the desired consistency. The clay is then pressed into steel moulds with a hydraulic press. The shaped clay is then fired ("burned") at 900–1000 °C to achieve strength.

### Rail Kilns

In modern brickworks, this is usually done in a continuously fired tunnel kiln, in which the bricks are fired as they move slowly through the kiln on conveyors, rails, or kiln cars, which achieves a more consistent brick product. The bricks often have lime, ash, and organic matter added, which accelerates the burning process.

### Bull's Trench Kilns

In India, brick making is typically a manual process. The most common type of brick kiln in use there is the Bull's Trench Kiln (BTK), based on a design developed by British engineer W. Bull in the late 19th century. An oval or circular trench is dug, 6–9 metres wide, 2-2.5 metres deep, and 100–150 metres in circumference. A tall exhaust chimney is constructed in the centre. Half or more of the trench is filled with "green" (unfired) bricks which are stacked in an open lattice pattern to allow airflow. The lattice is capped with a roofing layer of finished brick.

In operation, new green bricks, along with roofing bricks, are stacked at one end of the brick pile; cooled finished bricks are removed from the other end for transport to their destinations. In the middle, the brick workers create a firing zone by dropping fuel (coal, wood, oil, debris, and so on) through access holes in the roof above the trench.

The advantage of the BTK design is a much greater energy efficiency compared with clamp or scove kilns. Sheet metal or boards are used to route the airflow through the brick lattice so that fresh air flows first through the recently burned bricks, heating the air, then through the active burning zone. The air continues through the green brick zone (pre-heating and drying the bricks), and finally out the chimney, where the rising gases create suction which pulls air through the system. The reuse of heated air yields savings in fuel cost.

As with the rail process above, the BTK process is continuous. A half dozen labourers working around the clock can fire approximately 15,000–25,000 bricks a day. Unlike the rail process, in the BTK process the bricks do not move. Instead, the locations at which the bricks are loaded, fired, and unloaded gradually rotate through the trench.

***Figure:*** *West face of Roskilde Cathedral in Roskilde, Denmark.*

### Dry Pressed Bricks

The dry press method is similar to the soft mud brick method, but starts with a much thicker clay mix, so it forms more accurate, sharper-edged bricks. The greater force in pressing and the longer burn make this method more expensive.

### Extruded Bricks

For extruded bricks the clay is mixed with 10–15% water (stiff extrusion) or 20–25% water (soft extrusion). This mixture is forced through a die to create a long cable of material of the desired width and depth. This mass is then cut into bricks of the desired length by a wall of wires. Most structural bricks are made by this method as it produces hard, dense bricks, and suitable dies can produce holes or other perforations as well. The introduction of such holes reduces the volume of clay needed, and hence the cost. Hollow bricks are lighter and easier to handle, and have different thermal properties than solid bricks. The cut bricks are hardened by drying for 20 to 40 hours at 50 to 150 °C before being fired. The heat for drying is often waste heat from the kiln.

European-style extruded bricks or blocks are used in single-wall construction with finishes applied on the inside and outside. Their many voids comprise a greater proportion of the volume than the solid, thin walls of fired clay. Such bricks are made in 15-, 25-, 30-, 42- and 50-cm widths. Some models have very high thermal insulation properties, making them suitable for zero-energy buildings.

### Calcium-Silicate Bricks

The raw materials for calcium-silicate bricks include lime mixed with quartz, crushed flint or crushed siliceous rock together with mineral colourants. The materials are mixed and left until the lime is completely hydrated; the mixture is then pressed into moulds and cured in an autoclave for two or three hours to speed the chemical hardening. The finished bricks are very accurate and uniform, although the sharp arrises need careful handling to avoid damage to brick (and bricklayer). The bricks can be made in a variety of colours; white is common but pastel shades can be achieved. This type of brick is common in Sweden, especially in houses built or renovated in the 1970s, where it is known as "Mexitegel" (en: Mexi[can] Bricks). In India these are known as Fly ash bricks, manufactured using the FaL-G (fly ash, lime and gypsum) process. Calcium-silicate bricks are also manufactured in Canada and the United States, and meet the criteria set forth in ASTM C73 – 10 Standard Specification for Calcium Silicate Brick (Sand-Lime Brick). It has lower embodied energy than cement based man-made stone and clay brick.

### Concrete Bricks

Bricks of concrete with sand aggregate can be made using a simple machine and a basic assembly line. A conveyor belt adds the mixture to a machine, which pours a measured amount of concrete into a form. The form is vibrated to remove bubbles, after which it is raised to reveal the wet bricks, spaced out on a plywood sheet. A small elevator then stacks these palettes, after which a forklift operator moves them to the brickyard for drying.

### Influence on Fired Colour

The fired colour of clay bricks is influenced by the chemical and mineral content of the raw materials, the firing temperature, and the atmosphere in the kiln. For example, pink coloured bricks are the result of a high iron content, white or yellow bricks have a higher lime content. Most bricks burn to various red hues; as the temperature is increased the colour moves through dark red, purple and then to brown or grey at around 1,300 °C (2,372 °F). Calcium silicate bricks have a wider range of shades and colours, depending on the colourants used. The names of bricks may reflect their origin and colour, such as London stock brick and Cambridgeshire White.

"Bricks" formed from concrete are usually termed blocks, and are typically pale grey in colour. They are made from a dry, small aggregate concrete which is formed in steel moulds by vibration and compaction in either an "egglayer" or static machine. The finished blocks are cured rather than fired using low-pressure steam. Concrete blocks are manufactured in a much wider range of shapes and sizes than clay bricks and are also available with a wider range of face treatments – a number of which simulate the appearance of clay bricks.

An impervious and ornamental surface may be laid on brick either by salt glazing, in which salt is added during the burning process, or by the use of a "slip," which is a glaze material into which the bricks are dipped. Subsequent reheating in the kiln fuses the slip into a glazed surface integral with the brick base.

Natural stone bricks are of limited modern utility due to their enormous comparative mass, the consequent foundation needs, and the time-consuming and skilled labour needed in their construction and laying. They are very durable and considered more handsome than clay bricks by some. Only a few stones are suitable for bricks. Common materials are granite, limestone and sandstone. Other stones may be used (for example, marble, slate, quartzite, and so on) but these tend to be limited to a particular locality.

### Optimal Dimensions, Characteristics, and Strength

For efficient handling and laying, bricks must be small enough and light enough to be picked up by the bricklayer using one hand (leaving the other hand free for the trowel). Bricks are usually laid flat and as a result the effective limit on the width of a brick is set by the distance which can conveniently be spanned between the thumb and fingers of one hand, normally about four inches (about 100 mm). In most cases, the length of a brick is about twice its width, about eight inches (about 200 mm) or slightly more. This allows bricks to be laid *bonded* in a structure which increases stability and strength (for an example, see the illustration of bricks laid in *English bond,* at the head of this article). The wall is built using alternating courses of *stretchers*, bricks laid longways, and *headers*, bricks laid crossways. The headers tie the wall together over its width. In fact, this wall is built in a variation of *English bond* called *English cross bond* where the successive layers of stretchers are displaced horizontally from each other by half a brick length. In true *English bond*, the perpendicular lines of the stretcher courses are in line with each other.

A bigger brick makes for a thicker (and thus more insulating) wall. Historically, this meant that bigger bricks were necessary in colder climates, while a smaller brick was adequate, and more economical, in warmer regions. A notable illustration of this correlation is the Green Gate in Gdansk; built in 1571 of imported Dutch brick, too small for the colder climate of Gdansk, it was notorious for being a chilly and drafty residence. Nowadays this is no longer an issue, as modern walls typically incorporate specialised insulation materials.

The correct brick for a job can be selected from a choice of colour, surface texture, density, weight, absorption and pore structure, thermal characteristics, thermal and moisture movement, and fire resistance.

***Table:*** *Face brick ("house brick") sizes, (alphabetical order)*

| ***Standard*** | ***Imperial*** | ***Metric*** |
|---|---|---|
| Australia | 9 × 4⅓ × 3 in | 230 × 110 × 76 mm |
| Denmark | 9 × 4¼ × 2¼ in | 228 × 108 × 54 mm |
| Germany | 9 × 4¼ × 2¾ in | 240 × 115 × 71 mm |
| India | 9 × 4¼ × 2¾ in | 228 × 107 × 69 mm |
| Romania | 9 × 4¼ × 2½ in | 240 × 115 × 63 mm |
| Russia | 10 × 4¾ × 2½ in | 250 × 120 × 65 mm |
| South Africa | 8¾ × 4 × 3 in | 222 × 106 × 73 mm |
| Sweden | 10 × 4¾ × 2½ in | 250 × 120 × 62 mm |
| United Kingdom | 8½ × 4 × 2½ in | 215 × 102.5 × 65 mm |
| United States | 7] ⅝ × 3] ⅝ × 2¼ in | 194 × 92 × 57 mm |

In England, the length and width of the common brick has remained fairly constant over the centuries, but the depth has varied from about two inches (about 51 mm) or smaller in earlier times to about two and a half inches (about 64 mm) more recently. In the United Kingdom, the usual size of a modern brick is 215 × 102.5 × 65 mm (about 8 ⅝ × 4 × 2 ⅝ inches), which, with a nominal 10 mm (⅜ inch) mortar joint, forms a *unit size* of 225 × 112.5 × 75 mm (9 × 4 ½ × 3 inches), for a ratio of 6:3:2.

In the United States, modern standard bricks are (controlled by American Society for Testing and Materials ASTM) about 8 × 3 ⅝ ×2 ¼ inches (203 × 92 × 57 mm). The more commonly used is the modular brick 7 ⅝ ×3 ⅝ × 2 ¼ inches (194 × 92 × 57 mm). This modular brick of 7 ⅝ plus a ⅜ mortar joint eased the calculations of the number of bricks in a given run.

Some brickmakers create innovative sizes and shapes for bricks used for plastering (and therefore not visible) where their inherent mechanical properties are more important than their visual ones. These bricks are usually slightly larger, but not as large as blocks and offer the following advantages:

- a slightly larger brick requires less mortar and handling (fewer bricks), which reduces cost
- their ribbed exterior aids plastering
- more complex interior cavities allow improved insulation, while maintaining strength.

Blocks have a much greater range of sizes. Standard coordinating sizes in length and height (in mm) include 400×200, 450×150, 450×200, 450×225, 450×300, 600×150, 600×200, and 600×225; depths (work size, mm) include 60, 75, 90, 100, 115, 140, 150, 190, 200, 225, and 250. They are usable across this range as they are lighter than clay bricks. The density of solid clay bricks is around 2,000 kg/m$^3$: this is reduced by frogging, hollow bricks, and so on, but aerated autoclaved concrete, even as a solid brick, can have densities in the range of 450–850 kg/m$^3$.

Bricks may also be classified as *solid* (less than 25% perforations by volume, although the brick may be "frogged," having indentations on one of the longer faces), *perforated* (containing a pattern of small holes through the brick, removing no more than 25% of the volume), *cellular* (containing a pattern of holes removing more than 20% of the volume, but closed on one face), or *hollow* (containing a pattern of large holes removing more than 25% of the brick's volume). Blocks may be solid, cellular or hollow

The term "frog" for the indentation on one bed of the brick is a word that often excites curiosity as to its origin. The most likely explanation is that brickmakers also call the block that is placed in the mould to form the indentation a frog. Modern brickmakers usually use plastic frogs but in the past they were made of wood. When these are wet and have clay on them they resemble the amphibious kind of frog and this is where they got their name. Over time this term also came to refer to the indentation left by them.

The compressive strength of bricks produced in the United States ranges from about 1000 lbf/in$^2$ to 15,000 lbf/in$^2$ (7 to 105 MPa or N/mm$^2$ ), varying according to the use to which the brick are to be put. In England clay bricks can have strengths of up to 100 MPa, although a common house brick is likely to show a range of 20–40 MPa.

## *Use*

Bricks are used for building, block paving and pavement. In the USA, brick pavement was found incapable of withstanding heavy traffic, but it is coming back into use as a method of traffic calming or as a decorative surface in pedestrian precincts. For example, in the early 1900s, most of the streets in the city of Grand Rapids, Michigan were paved with brick. Today, there are only about 20 blocks of brick paved streets remaining (totalling less than 0.5 percent of all the streets in the city limits).

Bricks in the metallurgy and glass industries are often used for lining furnaces, in particular refractory bricks such as silica, magnesia, chamotte and neutral (chromomagnesite) refractory bricks. This type of brick must have good thermal shock resistance, refractoriness under load, high melting point, and satisfactory porosity. There is a large refractory brick industry, especially in the United Kingdom, Japan, the United States, Belgium and the Netherlands.

In Northwest Europe, bricks have been used in construction for centuries. Until recently, almost all houses were built almost entirely from bricks. Although many houses are now built using a mixture of concrete blocks and other materials, many houses are skinned with a layer of bricks on the outside for aesthetic appeal.

Engineering bricks are used where strength, low water porosity or acid (flue gas) resistance are needed.

In the UK a redbrick university is one founded and built in the Victorian era, often as a technical college. The term serves to distinguish these polytechnic colleges from older, more classics-oriented universities.

Colombian architect Rogelio Salmona was noted for his extensive use of red brick in his buildings and for using natural shapes like spirals, radial geometry and curves in his designs. Most buildings in Colombia are made of brick, given the abundance of clay in equatorial countries like this one.

### *Limitations*

Starting in the 20th century, the use of brickwork declined in some areas due to concerns with earthquakes. Earthquakes such as the San Francisco earthquake of 1906 and the 1933 Long Beach eathquake revealed the weaknesses of brick masonry in earthquake-prone areas. During seismic events, the mortar cracks and crumbles, and the bricks are no longer held together. Brick masonry with steel reinforcement, which helps hold the masonry together during earthquakes, was used to replace many of the unreinforced masonry buildings. Retrofitting older unreinforced masonry structures has been mandated in many jurisdictions.

## Gypsum

Gypsum is a very soft sulfate mineral composed of calcium sulfatedihydrate, with the chemical formula $CaSO_4 \cdot 2H_2O$. It can be used as a fertilizer, is the main constituent in many forms of plaster and is widely mined. A very fine-grained white or lightly tinted variety of gypsum, called alabaster, has been used for sculpture by many cultures including Ancient Egypt, Mesopotamia and the Nottingham alabasters of medieval England. It is the definition of a hardness of 2 on the Mohs scale of mineral hardness. It forms as an evaporite mineral and as a hydration product of anhydrite.

### *Etymology and History*

The word gypsum is derived from the Greek word ãýøïò (*gypsos*), "chalk" or "plaster". Because gypsum from the quarries of the Montmartre district of Paris have long furnished burnt gypsum (calcined gypsum) used for various purposes, this dehydrated gypsum became known as plaster of Paris. Upon addition of water, after a few tens of minutes plaster of Paris becomes regular gypsum (dihydrate) again, causing the material to harden or "set" in ways that are useful for casting and construction.

Gypsum was known in Old English as *sp¿rstân*, "spear stone", referring to its crystalline projections. (Thus, the word spar in mineralogy is by way of comparison to gypsum, referring to any non-ore mineral or crystal that forms in spearlike projections.) Gypsum may act as a source of sulfur for plant growth, and in the early 19th century, it was regarded as an almost miraculous fertilizer. American farmers were so anxious to acquire it that a lively smuggling trade with Nova Scotia evolved, resulting in the so-called "Plaster War" of 1812.

### Physical Properties

Gypsum is moderately water-soluble (~2.0–2.5 g/l at 25°C) and, in contrast to most other salts, it exhibits a retrograde solubility, becoming less soluble at higher temperatures. When gypsum is heated in air it loses water and converts first to calcium sulfate hemihydrate, (bassanite, often simply called "plaster") and, if heated further, to anhydrous calcium sulfate (anhydrite). As for anhydrite, its solubility in saline solutions and in brines is also strongly dependent on NaCl concentration.

Gypsum crystals are found to contain anion water and hydrogen bonding.

### Crystal Varieties

Gypsum occurs in nature as flattened and often twinned crystals, and transparent, cleavable masses called selenite. Selenite contains no significant selenium; rather, both substances were named for the ancient Greek word for the Moon.

Selenite may also occur in a silky, fibrous form, in which case it is commonly called "satin spar". Finally, it may also be granular or quite compact. In hand-sized samples, it can be anywhere from transparent to opaque. A very fine-grained white or lightly tinted variety of gypsum, called alabaster, is prized for ornamental work of various sorts. In arid areas, gypsum can occur in a flower-like form, typically opaque, with embedded sand grains called desert rose. It also forms some of the largest crystals found in nature, up to 12 metres (39 ft) long, in the form of selenite.

### Occurrence

Gypsum is a common mineral, with thick and extensive evaporite beds in association with sedimentary rocks. Deposits are known to occur in strata from as far back as the Archaeaneon. Gypsum is deposited from lake and sea water, as well as in hot springs, from volcanic vapors, and sulfate solutions in veins. Hydrothermalan hydrite in veins is commonly hydrated to gypsum by groundwater in near-surface exposures. It is often associated with the minerals halite and sulfur. Pure gypsum is white, but other substances found as impurities may give a wide range of colours to local deposits.

Because gypsum dissolves over time in water, gypsum is rarely found in the form of sand. However, the unique conditions of the White Sands National Monument in the US state of New Mexico have created a 710 km$^2$ (270 sq mi) expanse of white gypsum sand, enough to supply the construction industry with drywall for 1,000 years. Commercial exploitation of the area, strongly opposed by area residents, was permanently prevented in 1933 when president Herbert Hoover declared the gypsum dunes a protected national monument.

***Figure:*** *Veins of gypsum in the silts/marls of the Tea Green and Grey Marls, Blue Anchor, Somerset, UK.*

Gypsum is also formed as a by-product of sulfideoxidation, amongst others by pyriteoxidation, when the sulfuric acid generated reacts with calcium carbonate. Its presence indicates oxidizing conditions. Under reducing conditions, the sulfates it contains can be reduced back to sulfide by sulfate reducing bacteria. Electric power stations burning coal with flue gas desulfurization produce large quantities of gypsum as a byproduct from the scrubbers.

Orbital pictures from the Mars Reconnaissance Orbiter (MRO) have indicated the existence of gypsum dunes in the northern polar region of Mars, which were later confirmed at ground level by the Mars Exploration Rover (MER) *Opportunity*.

### Mining

Commercial quantities of gypsum are found in the cities of Araripina and Grajaú, Brazil, Pakistan, Jamaica, Iran (world's second largest producer), Thailand, Spain (the main producer in Europe), Germany, Italy, England, Ireland, in British Columbia, Manitoba, Ontario, Nova Scotia and Newfoundland in Canada, and in New York, Michigan, Indiana, Texas (in the Palo Duro Canyon), Iowa, Kansas,

Oklahoma, Arizona, New Mexico, Colorado, Utah, Arkansas and Nevada in the United States.

There is also a large open pit quarry located at Plaster City, California, in Imperial County, and in East Kutai, Kalimantan. Several small mines also exist in places such as Kalannie in Western Australia, where gypsum is sold to private buyers for changing the pH levels of soil for agricultural purposes.

***Figure:*** *Golden gypsum crystals from Winnipeg.*

Crystals of gypsum up to 11 metres (36 ft) long have been found in the caves of the Naica Mine of Chihuahua, Mexico. The crystals thrived in the cave's extremely rare and stable natural environment. Temperatures stayed at 58°C (136°F), and the cave was filled with mineral-rich water that drove the crystals' growth. The largest of those crystals weighs 55 short tons (50,000 kg) and is around 500,000 years old.

### *Synthesis*

Synthetic gypsum is recovered via flue-gas desulfurization at some coal-fired power plants. It can be used interchangeably with natural gypsum in some applications.

Gypsum also precipitates onto brackish water membranes, a phenomenon known as mineral salt scaling, such as during brackish water desalination of water with high concentrations of calcium and sulfate. Scaling decreases membrane life and productivity.

This is one of the main obstacles in brackish water membrane desalination processes, such as reverse osmosis or nanofiltration. Other forms of scaling, such as calcite scaling, depending on the water source, can also be important considerations in distillation, as well as in heat exchangers, where either the salt solubility or concentration can change rapidly.

A new study has suggested that the formation of gypsum starts as tiny crystals of a mineral called bassanite ($CaSO_4 \cdot 0.5H_2O$). This process occurs via a three-stage pathway: (1) homogeneous nucleation of nanocrystalline bassanite; (2) self-assembly of bassanite into aggregates, and (3) transformation of bassanite into gypsum.

### *Uses of Gypsum*

Gypsum is used in a wide variety of applications:

- Gypsum board is primarily used as a finish for walls and ceilings, and is known in construction as drywall or plasterboard.
- Plaster ingredient (surgical splints, casting moulds, modelling)
- Fertilizer and soil conditioner: In the late 18th and early 19th centuries, Nova Scotia gypsum, often referred to as plaster, was a highly sought fertilizer for wheat fields in the United States. It is also used in ameliorating high sodium soils.
- A binder in fast-dry tennis court clay
- As alabaster, a material for sculpture, especially in the ancient world before steel was developed, when its relative softness made it much easier to carve than stone with available tools.
- A wood substitute in the ancient world: For example, when wood became scarce due to deforestation on Bronze AgeCrete, gypsum was employed in building construction at locations where wood was previously used.

- A tofu (soy bean curd) coagulant, making it ultimately a major source of dietary calcium, especially in Asian cultures which traditionally use few dairy products
- Adding hardness to water used for homebrewing
- Used in baking as a dough conditioner, reducing stickiness, and as a baked-goods source of dietary calcium. The primary component of mineral yeast food.
- A component of Portland cement used to prevent flash setting of concrete
- Soil/water potential monitoring (soil moisture tension)
- A common ingredient in making mead
- In the medieval period, it was mixed, by scribes and illuminators, with lead carbonate (powdered white lead) to make gesso, which was applied to illuminated letters and gilded with gold in illuminated manuscripts.
- In foot creams, shampoos and many other hair products
- A medicinal agent in traditional Chinese medicine called *shigao*
- Impression plasters in dentistry
- Used in mushroom cultivation to stop grains from clumping together

# 6

# Use of Pozzolan

A pozzolan is a siliceous or siliceous and aluminous material which, in itself, possesses little or no cementitious value but which will, in finely divided form and in the presence of water, react chemically with calcium hydroxide at ordinary temperature to form compounds possessing cementitious properties (ASTM C618). The broad definition of a pozzolan imparts no bearing on the origin of the material, only on its capability of reacting with calcium hydroxide and water. A quantification of this capability is comprised in the term pozzolanic activity.

## History

Mixtures of calcined lime and finely ground reactive (alumino-)silicate materials were pioneered and developed as inorganic binders in the Antique world. Architectural remains of the Minoan civilization on Crete have shown evidence of the combined use of slaked lime and additions of finely ground potsherds for water-proof renderings in baths, cisterns and aqueducts. Evidence of the deliberate use of volcanic materials such as volcanic ashes or tuffs by the ancient Greeks dates back to at least 500-400 BC, as uncovered at the ancient city of Kameiros, Rhodes. In subsequent centuries the practice spread to the mainland and was eventually adopted and further developed by the Romans.

The Romans used volcanic pumices and tuffs found in neighbouring territories, the most famous ones found in Pozzuoli (Naples), hence the name pozzolan, and in Segni (Latium). Preference was given to natural pozzolan sources (such as German Trass), but crushed ceramic waste was frequently used when natural deposits were not locally available. The exceptional lifetime and preservation conditions of some of the most famous Roman buildings such as the Pantheon or the Pont du Gard constructed using pozzolan-lime mortars and concrete testify to both the excellent workmanship reached by Roman "engineers" and to the durable properties of the utilised binders.

Much of the practical skills and knowledge regarding the use of pozzolans was lost at the decline of the Roman empire. The rediscovery of Roman architectural practices as described by Vitruvius in *De architectura*, also led to the reintroduction of lime-pozzolan binders. Particularly the strength, durability and hydraulic capability of hardening underwater made them popular construction materials during the 16th-18th century. The invention of hydraulic cements and eventually Portland cement in the 18th and 19th century resulted in a gradual decline of the use of pozzolan-lime binders, which develop strength less rapidly.

Over the course of the 20th century the use of pozzolans as addition (the technical term is "supplementary cementitious material", usually abbreviated "SCM") to Portland cement concrete mixtures has become common practice. Combinations of economical and technical aspects and, increasingly, environmental concerns have made so-called blended cements, i.e. cements that contain considerable amounts of supplementary cementitious materials (mostly around 20 wt.%, but over 80 wt.% in Portland blast-furnace slag cement) the most widely produced and used cement type by the beginning of the 21st century.

### Pozzolanic Materials

The general definition of a pozzolan embraces a large number of materials which vary widely in terms of origin, composition and properties. Both natural and artificial (man-made) materials show pozzolanic activity and are used as supplementary cementitious materials. Artificial pozzolans can be produced deliberately, for instance by thermal activation of kaolin-clays to obtain metakaolin, or can be obtained as waste or by-products from high-temperature process such as fly ashes from coal-fired electricity production. The most commonly used pozzolans today are industrial by-products such as fly ash, silica fume from silicon smelting, highly reactive metakaolin, and burned organic matter residues rich in silica such as rice husk ash. Their use has been firmly established and regulated in many countries. However, the supply of high-quality pozzolanic by-products is limited and many local sources are already fully exploited. Alternatives to the established pozzolanic by-products are to be found on the one hand in an expansion of the range of industrial by-products or societal waste considered and on the other hand in an increased usage of naturally occurring pozzolans.

Natural pozzolans are abundant in certain locations and are extensively used as an addition to Portland cement in countries such as for example Italy, Germany, Greece and China. The great majority of natural pozzolans in use today is of volcanic origin. Volcanic ashes and pumices largely composed of volcanic glass are commonly used, as are deposits in which the volcanic glass has been altered to zeolites by interaction with alkaline waters. Deposits of sedimentary origin are less common. Diatomaceous earths, formed by the accumulation of siliceous diatom microskeletons, are a prominent source material here.

### Use

The benefits of pozzolan utilisation in cement and concrete are threefold. First is the economic gain obtained by replacing a substantial part of the Portland cement by cheaper natural pozzolans or industrial by-products. Second is the lowering of the blended cement environmental cost associated with the greenhouse gases emitted during Portland cement production. A third advantage is the durability improvement of the end product. Additionally, the increased blending of pozzolans with Portland cement is of limited interference in the conventional production process and offers the opportunity to valorize large amounts of industrial and societal waste into durable construction materials.

Current practice may permit up to a 40 percent reduction of Portland cement used in the concrete mix when replaced with a carefully designed combination of approved pozzolans. When the mix is designed properly, concrete can utilise pozzolans without significantly reducing the final compressive strength or other performance characteristics. The properties of hardened blended cements are strongly related to the development of the binder microstructure, i.e., to the distribution, type, shape and dimensions of both reaction products and pores. The beneficial effects of pozzolan addition in terms of compressive strength performance and durability are mostly attributed to the pozzolanic reaction in which calcium hydroxide is consumed to produce additional C-S-H and C-A-H reaction products. These pozzolanic reaction products fill in pores and result in a refining of the pore size distribution or pore structure. This results in a lowered permeability of the binder.

The contribution of the pozzolanic reaction to cement strength is usually developed at later curing stages, depending on the pozzolanic activity. In the large majority of blended cements initial lower strengths can be observed compared to the parent Portland cement. However, especially in the case of pozzolans finer than the Portland cement, the decrease in early strength is usually less than what can be expected based on the dilution factor. This can be explained by the filler effect, in which small SCM grains fill in the space between the cement particles, resulting in a much denser binder. The acceleration of the Portland cement hydration reactions can also partially accommodate the loss of early strength.

The increased chemical resistance to the ingress and harmful action of aggressive solutions constitutes one of the main advantages of pozzolan blended cements. The improved durability of the pozzolan-blended binders enables to lengthen the service life of structures and reduces the costly and inconvenient need to replace damaged constructions. One of the principal reasons increased durability is the lowered calcium hydroxide content available to take part in deleterious expansive reactions induced by e.g. sulfate attack. Furthermore, the reduced binder permeability slows down the ingress of harmful ions such as chlorine or carbonate. The pozzolanic reaction can also reduce the risk of expansive alkali-silica reactions between the cement and aggregates by changing the binder pore solution. Lowering the solution alkalinity and increasing alumina concentrations strongly decreases or inhibits the dissolution of the aggregate aluminosilicates.

## Lime (Material)

Lime is a general term for calcium-containing inorganic materials, in which carbonates, oxides and hydroxides predominate. Strictly speaking, lime is calcium oxide or calcium hydroxide. It is also the name of the natural mineral (native lime) CaO which occurs as a product of coal seam fires and in altered limestonexenoliths in volcanicejecta. The word “lime” originates with its earliest use as building mortar and has the sense of “sticking or adhering.”

These materials are still used in large quantities as building and engineering materials (including limestone products, concrete and mortar) and as chemical feedstocks, among other uses. Lime industries and the use of many of the resulting products date from prehistoric periods in both the Old World and the New World. Lime is used extensively for waste water treatment with ferrous sulfate.

The rocks and minerals from which these materials are derived, typically limestone or chalk, are composed primarily of calcium carbonate. They may be cut, crushed or pulverized and chemically altered. "Burning" (calcination) converts them into the highly caustic material *quicklime* (calcium oxide, CaO) and, through subsequent addition of water, into the less caustic (but still strongly alkaline) *slaked lime* or *hydrated lime* (calcium hydroxide, $Ca(OH)_2$), the process of which is called *slaking of lime.*

When the term is encountered in an agricultural context, it probably refers to agricultural lime. Otherwise it most commonly means slaked lime, as the more dangerous form is usually described more specifically as quicklime or *burnt lime.*

### Lime Production Process

Limestone is extracted from quarries or mines. Part of the extracted stone, selected according to its chemical composition and granulometry, is calcinated at about 1000 °C in different types of lime kiln.

Quicklime is produced according to the reaction: $CaCO_3$ + heat $\rightarrow$ CaO + $CO_2$.

Quicklime can be hydrated, i.e., combined with water. Hydrated lime, known as slaked lime, is produced according to the reaction: CaO + $H_2O \rightarrow Ca(OH)_2$.

### The Lime Cycle

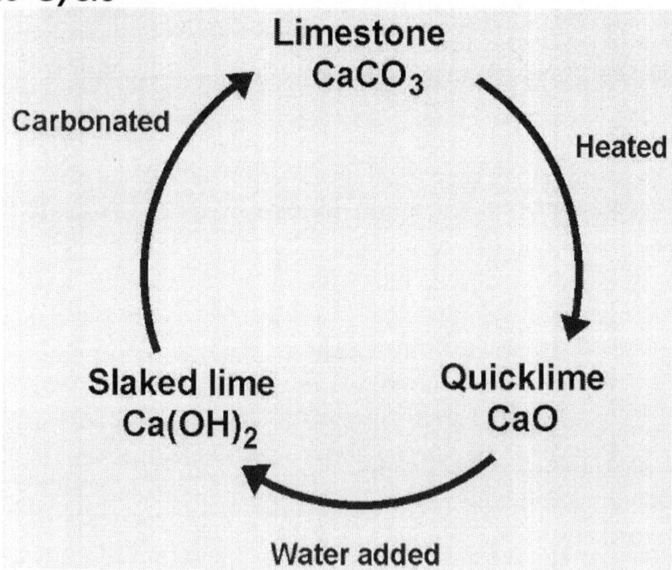


***Figure:*** *The_lime_cycle.jpg.*

Slaked lime (calcium hydroxide) is mixed into a thick slurry with sand and water to form various kinds of mortar and render for building purposes. Lime mortar was traditionally used in the joints between bricks or stones in masonry building construction. Now cement is usually added to the mixture to form a harder mortar. When the masonry has been laid, the slaked lime in the mortar slowly begins to react with carbon dioxide to form calcium carbonate (limestone) according to the reaction: $Ca(OH)_2 + CO_2 \rightarrow CaCO_3 + H_2O$. The carbon dioxide that takes part in this reaction is principally available dissolved in rainwater rather than in gaseous form.

This process by which limestone is converted to quicklime by heating, then to slaked lime by hydration, and reverted to limestone by carbonation is known as the Lime Cycle.

### *Lime in Roman Concrete*

The Romans made concrete by mixing lime and volcanic rock. For underwater structures, lime and volcanic ash were mixed to form mortar, and this mortar and volcanic tuff were packed into wooden forms. The seawater instantly triggered an exothermic chemical reaction. The lime was hydrated – incorporating water molecules into its structure – and reacted with the ash to cement the whole mixture together.

## Polymer

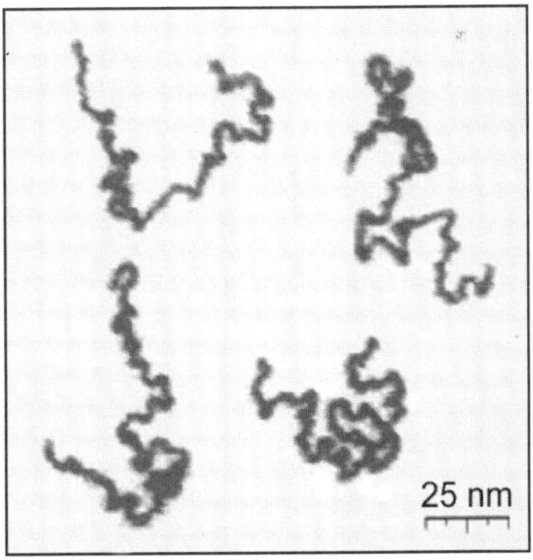


***Figure:*** *Appearance of real linear polymer chains as recorded using an atomic force microscope on a surface, under liquid medium. Chain contour length for this polymer is ~204 nm; thickness is ~0.4 nm.*

### IUPAC Definition

Substance composed of macromolecules.

***Note***: Applicable to substance macromolecular in nature like cross-linked

systems that can be considered as one macromolecule.

A polymer/ÈÙpRÙlhÙmYÙr/ is a large molecule composed of many repeated subunits, known as monomers. Because of their broad range of properties, both synthetic and natural polymers play an essential and ubiquitous role in everyday life. Polymers range from familiar synthetic plastics such as polystyrene (or styrofoam) to natural biopolymers such as DNA and proteins that are fundamental to biological structure and function. Polymers, both natural and synthetic, are created via polymerization of many monomers. Their consequently large molecular mass relative to small molecule compounds produces unique physical properties, including toughness, viscoelasticity, and a tendency to form glasses and semicrystalline structures rather than crystals.

The term "polymer" derives from the ancient Greek word ðïëýò (polus, meaning "many, much") and ìÝñïò (meros, meaning "parts"), and refers to a molecule whose structure is composed of multiple repeating units, from which originates a characteristic of high relative molecular mass and attendant properties. The units composing polymers derive, actually or conceptually, from molecules of low relative molecular mass. The term was coined in 1833 by Jöns Jacob Berzelius, though with a definition distinct from the modern IUPAC definition.

Polymers are studied in the fields of biophysics and macromolecular science, and polymer science (which includes polymer chemistry and polymer physics). Historically, products arising from the linkage of repeating units by covalent chemical bonds have been the primary focus of polymer science; emerging important areas of the science now focus on non-covalent links. Polyisoprene of latex rubber and the polystyrene of styrofoam are examples of polymeric natural/biological and synthetic polymers, respectively. In biological contexts, essentially all biological macromolecules—i.e., proteins (polyamides), nucleic acids (polynucleotides), and polysaccharides—are purely polymeric, or are composed in large part of polymeric components—e.g., isoprenylated/ lipid-modified glycoproteins, where small lipidic molecule and oligosaccharide modifications occur on the polyamide backbone of the protein.

### Common Examples

Natural polymeric materials such as shellac, amber, wool, silk and natural rubber have been used for centuries. A variety of other natural polymers exist, such as cellulose, which is the main constituent of wood and paper. The list of synthetic polymers includes synthetic rubber, phenol formaldehyde resin (or Bakelite), neoprene, nylon, polyvinyl chloride (PVC or vinyl), polystyrene, polyethylene, polypropylene, polyacrylonitrile, PVB, silicone, and many more.

Most commonly, the continuously linked backbone of a polymer used for the preparation of plastics consists mainly of carbon atoms. A simple example is polyethylene ('polythene' in British English), whose repeating unit is based on ethylenemonomer. However, other structures do exist; for example, elements such as silicon form familiar materials such as silicones, examples being Silly Putty and waterproof plumbing sealant. Oxygen is also commonly present in polymer backbones, such as those of polyethylene glycol, polysaccharides (in glycosidic bonds), and DNA (in phosphodiester bonds).

### Polymer Synthesis

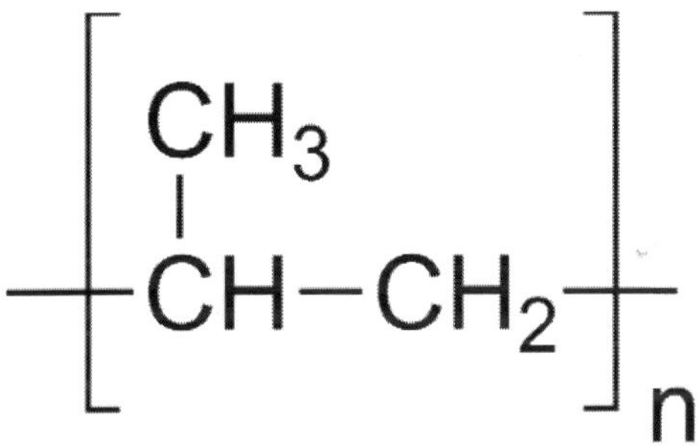

***Figure:*** *The repeating unit of the polymer polypropylene*

Polymerization is the process of combining many small molecules known as monomers into a covalently bonded chain or network. During the polymerization process, some chemical groups may be lost from each monomer. This is the case, for example, in the polymerization of PET polyester. The monomers are terephthalic acid ($HOOC-C_6H_4-COOH$) and ethylene glycol ($HO-CH_2-CH_2-OH$) but the repeating unit is $-OC-C_6H_4-COO-CH_2-CH_2-O-$, which corresponds to the combination of the two monomers with the loss of two water molecules. The distinct piece of each monomer that is incorporated into the polymer is known as a repeat unit or monomer residue.

Laboratory synthetic methods are generally divided into two categories, step-growth polymerization and chain-growth polymerization. The essential difference between the two is that in chain growth polymerization, monomers are added to the chain one at a time only, such as in polyethylene, whereas in step-growth polymerization chains of monomers may combine with one another directly, such as in polyester. However, some newer methods such as plasma polymerization do not fit neatly into either category. Synthetic polymerization reactions may be carried out with or without a catalyst. Laboratory synthesis of biopolymers, especially of proteins, is an area of intensive research.

### Biological Synthesis

There are three main classes of biopolymers: polysaccharides, polypeptides, and polynucleotides. In living cells, they may be synthesized by enzyme-mediated processes, such as the formation of DNA catalyzed by DNA polymerase. The synthesis of proteins involves multiple enzyme-mediated processes to transcribe genetic information from the DNA to RNA and subsequently translate that information to synthesize the specified protein from amino acids. The protein may be modified further following translation in order to provide appropriate structure and functioning.

### Modification of Natural Polymers

Naturally occurring polymers such as cotton, starch and rubber were familiar materials for years before synthetic polymers such as polyethene and perspex appeared on the market. Many commercially important polymers are synthesized by chemical modification of naturally occurring polymers. Prominent examples include the reaction of nitric acid and cellulose to form nitrocellulose and the formation of vulcanized rubber by heating natural rubber in the presence of sulfur. Ways in which polymers can be modified include oxidation, cross-linking and end-capping.

Especially in the production of polymers, the gas separation by membranes has acquired increasing importance in the petrochemical industry and is now a relatively well-established unit operation. The process of polymer degassing is necessary to suit polymer for extrusion and pelletizing, increasing safety, environmental, and product quality aspects. Nitrogen is generally used for this purpose, resulting in a vent gas primarily composed of monomers and nitrogen.

### Polymer Properties

Polymer properties are broadly divided into several classes based on the scale at which the property is defined as well as upon its physical basis. The most basic property of a polymer is the identity of its constituent monomers. A second set of properties, known as microstructure, essentially describe the arrangement of these monomers within the polymer at the scale of a single chain. These basic structural properties play a major role in determining bulk physical properties of the polymer, which describe how the polymer behaves as a continuous macroscopic material. Chemical properties, at the nano-scale, describe how the chains interact through various physical forces. At the macro-scale, they describe how the bulk polymer interacts with other chemicals and solvents.

### Monomers and Repeat Units

The identity of the monomer residues (repeat units) comprising a polymer is its first and most important attribute. Polymer nomenclature is generally based upon the type of monomer residues comprising the polymer. Polymers that contain only a single type of repeat unit are known as homopolymers, while polymers containing a mixture of repeat units are known as copolymers. Poly(styrene), for example, is composed only of styrene monomer residues, and is therefore classified as a homopolymer. Ethylene-vinyl acetate, on the other hand, contains more than one variety of repeat unit and is thus a copolymer. Some biological polymers are composed of a variety of different but structurally related monomer residues; for example, polynucleotides such as DNA are composed of a variety of nucleotide subunits. A polymer molecule containing ionizable subunits is known as a polyelectrolyte or ionomer.

### Microstructure

The microstructure of a polymer (sometimes called configuration) relates to the physical arrangement of monomer residues along the backbone of the chain. These are the elements of polymer structure that require the breaking of a covalent bond in order to change. Structure has a strong influence on the other properties of a polymer. For example, two samples of natural rubber may exhibit different durability, even though their molecules comprise the same monomers.

### Polymer Architecture

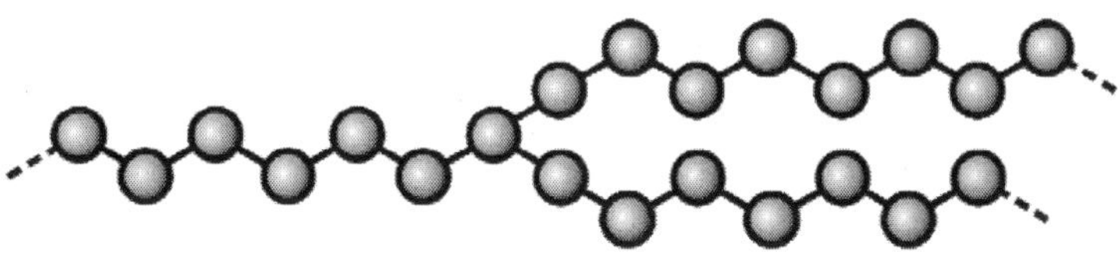

*Figure: Branch point in a polymer*

An important microstructural feature of a polymer is its architecture and shape, which relates to the way branch points lead to a deviation from a simple linear chain. A branched polymer molecule is composed of a main chain with one or more substituent side chains or branches. Types of branched polymers include star polymers, comb polymers, brush polymers, dendronized polymers, *ladders*, and dendrimers. There exist also two-dimensional polymers which are composed of topologically planar repeat units.

A polymer's architecture affects many of its physical properties including, but not limited to, solution viscosity, melt viscosity, solubility in various solvents, glass transition temperature and the size of individual polymer coils in solution.

A variety of techniques may be employed for the synthesis of a polymeric material with a range of architectures, for example Living polymerization.

## Chain Length

The physical properties of a polymer are strongly dependent on the size or length of the polymer chain. For example, as chain length is increased, melting and boiling temperatures increase quickly. Impact resistance also tends to increase with chain length, as does the viscosity, or resistance to flow, of the polymer in its melt state. Melt viscosity $\eta$is related to polymer chain length Z roughly as $\eta \sim Z_{22}$, so that a tenfold increase in polymer chain length results in a viscosity increase of over 1000 times.

Increasing chain length furthermore tends to decrease chain mobility, increase strength and toughness, and increase the glass transition temperature ($T_g$). This is a result of the increase in chain interactions such as Van der Waals attractions and entanglements that come with increased chain length. These interactions tend to fix the individual chains more strongly in position and resist deformations and matrix breakup, both at higher stresses and higher temperatures.

A common means of expressing the length of a chain is the degree of polymerization, which quantifies the number of monomers incorporated into the chain. As with other molecules, a polymer's size may also be expressed in terms of molecular weight. Since synthetic polymerization techniques typically yield a polymer product including a range of molecular weights, the weight is often expressed statistically to describe the distribution of chain lengths present in the same. Common examples are the number average molecular weight and weight average molecular weight. The ratio of these two values is the polydispersity index, commonly used to express the "width" of the molecular weight distribution. A final measurement is contour length, which can be understood as the length of the chain backbone in its fully extended state.

The flexibility of an unbranched chain polymer is characterized by its persistence length.

### *Monomer Arrangement in Copolymers*

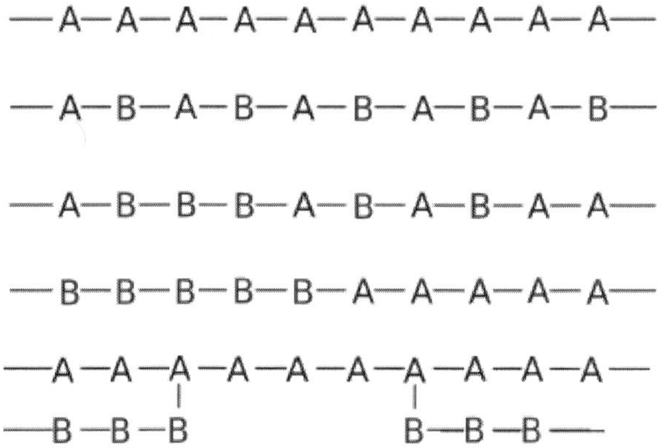


1 2 3 4 5

Monomers within a copolymer may be organised along the backbone in a variety of ways.

- Alternating copolymers possess regularly alternating monomer residues: $[AB...]_n$.
- Periodic copolymers have monomer residue types arranged in a repeating sequence: $[A_nB_m...]$ m being different from n .
- Statistical copolymers have monomer residues arranged according to a known statistical rule. A statistical copolymer in which the probability of finding a particular type of monomer residue at a particular point in the chain is independent of the types of surrounding monomer residue may be referred to as a truly random copolymer.
- Block copolymers have two or more homopolymer subunits linked by covalent bonds. Polymers with two or three blocks of two distinct chemical species (e.g., A and B) are called diblock copolymers and triblock copolymers, respectively. Polymers with three blocks, each of a different chemical species (e.g., A, B, and C) are termed triblockterpolymers.
- Graft or grafted copolymers contain side chains that have a different composition or configuration than the main chain.

## Tacticity

Tacticity describes the relative stereochemistry of chiralcentres in neighbouring structural units within a macromolecule. There are three types: isotactic (all substituents on the same side), atactic (random placement of substituents), and syndiotactic (alternating placement of substituents).

## Polymer Morphology

Polymer morphology generally describes the arrangement and microscale ordering of polymer chains in space.

## Crystallinity

When applied to polymers, the term *crystalline* has a somewhat ambiguous usage. In some cases, the term *crystalline* finds identical usage to that used in conventional crystallography. For example, the structure of a crystalline protein or polynucleotide, such as a sample prepared for x-ray crystallography, may be defined in terms of a conventional unit cell composed of one or more polymer molecules with cell dimensions of hundreds of angstroms or more.

A synthetic polymer may be loosely described as crystalline if it contains regions of three-dimensional ordering on atomic (rather than macromolecular) length scales, usually arising from intramolecular folding and/or stacking of adjacent chains. Synthetic polymers may consist of both crystalline and amorphous regions; the degree of crystallinity may be expressed in terms of a weight fraction or volume fraction of crystalline material. Few synthetic polymers are entirely crystalline.

The crystallinity of polymers is characterized by their degree of crystallinity, ranging from zero for a completely non-crystalline polymer to one for a theoretical completely crystalline polymer. Polymers with microcrystalline regions are generally tougher (can be bent more without breaking) and more impact-resistant than totally amorphous polymers. Polymers with a degree of crystallinity approaching zero or one will tend to be transparent, while polymers with intermediate degrees of crystallinity will tend to be opaque due to light scattering by crystalline or glassy regions. Thus for many polymers, reduced crystallinity may also be associated with increased transparency.

## Chain Conformation

The space occupied by a polymer molecule is generally expressed in terms of radius of gyration, which is an average distance from the centre of mass of the chain to the chain itself. Alternatively, it may be expressed in terms of pervaded volume, which is the volume of solution spanned by the polymer chain and scales with the cube of the radius of gyration.

### Mechanical Properties

The bulk properties of a polymer are those most often of end-use interest. These are the properties that dictate how the polymer actually behaves on a macroscopic scale.

### Tensile Strength

The tensile strength of a material quantifies how much stress the material will endure before suffering permanent deformation. This is very important in applications that rely upon a polymer's physical strength or durability. For example, a rubber band with a higher tensile strength will hold a greater weight before snapping. In general, tensile strength increases with polymer chain length and crosslinking of polymer chains.

### Young's Modulus of Elasticity

Young's Modulus quantifies the elasticity of the polymer. It is defined, for small strains, as the ratio of rate of change of stress to strain. Like tensile strength, this is highly relevant in polymer applications involving the physical properties of polymers, such as rubber bands. The modulus is strongly dependent on temperature. Viscoelasticity describes a complex time-dependent elastic response, which will exhibit hysteresis in the stress-strain curve when the load is removed. Dynamic mechanical analysis or DMA measures this complex modulus by oscillating the load and measuring the resulting strain as a function of time.

### Transport Properties

Transport properties such as diffusivity relate to how rapidly molecules move through the polymer matrix. These are very important in many applications of polymers for films and membranes.

### Phase Behaviour

***Melting Point:*** The term *melting point*, when applied to polymers, suggests not a solid-liquid phase transition but a transition from a crystalline or semi-crystalline phase to a solid amorphous phase. Though abbreviated as simply $T_m$, the property in question is more properly called the crystalline melting temperature. Among synthetic polymers, crystalline melting is only discussed with regards to thermoplastics, as thermosetting polymers will decompose at high temperatures rather than melt.

### Glass Transition Temperature

A parameter of particular interest in synthetic polymer manufacturing is the glass transition temperature ($T_g$), at which amorphous polymers undergo a transition from a rubbery, viscous liquid, to a brittle, glassy amorphous solid on cooling. The glass transition temperature may be engineered by altering the degree of branching or crosslinking in the polymer or by the addition of plasticizer.

### Mixing Behaviour

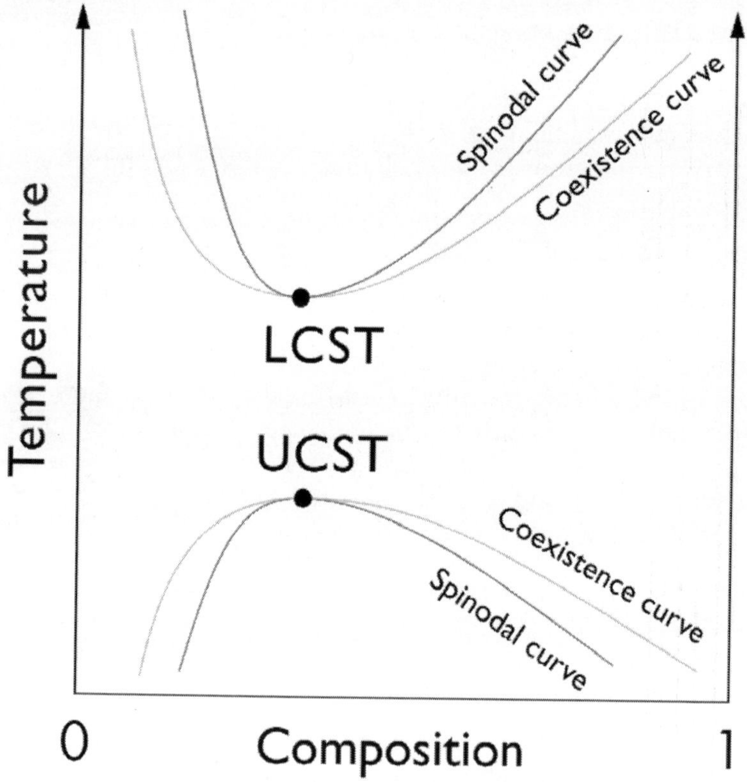


***Figure:*** *Phase diagram of the typical mixing behavior of weakly interacting polymer solutions.*

In general, polymeric mixtures are far less miscible than mixtures of small molecule materials. This effect results from the fact that the driving force for mixing is usually entropy, not interaction energy. In other words, miscible materials usually form a solution not because their interaction with each other is more favourable than their self-interaction, but because of an increase in entropy and hence free energy associated with increasing the amount of volume available to each component. This increase in entropy scales with the number of particles (or moles) being mixed. Since polymeric molecules are much larger and hence generally have much higher specific volumes than small molecules, the number of molecules involved in a polymeric mixture is far smaller than the number in a small molecule mixture of equal volume. The energetics of mixing, on the other hand, is comparable on a per volume basis for polymeric and small molecule mixtures. This tends to increase the free energy of mixing for polymer solutions and thus make solvation less favourable. Thus, concentrated solutions of polymers are far rarer than those of small molecules.

Furthermore, the phase behavior of polymer solutions and mixtures is more complex than that of small molecule mixtures. Whereas most small molecule solutions exhibit only an upper critical solution temperature phase transition, at which phase separation occurs with cooling, polymer mixtures commonly exhibit a lower critical solution temperature phase transition, at which phase separation occurs with heating.

In dilute solution, the properties of the polymer are characterized by the interaction between the solvent and the polymer. In a good solvent, the polymer appears swollen and occupies a large volume. In this scenario, intermolecular forces between the solvent and monomer subunits dominate over intramolecular interactions. In a bad solvent or poor solvent, intramolecular forces dominate and the chain contracts. In the theta solvent, or the state of the polymer solution where the value of the second virial coefficient becomes 0, the intermolecular polymer-solvent repulsion balances exactly the intramolecular monomer-monomer attraction. Under the theta condition (also called the Flory condition), the polymer behaves like an ideal random coil. The transition between the states is known as a coil-globule transition.

### Inclusion of Plasticizers

Inclusion of plasticizers tends to lower $T_g$ and increase polymer flexibility. Plasticizers are generally small molecules that are chemically similar to the polymer and create gaps between polymer chains for greater mobility and reduced interchain interactions. A good example of the action of plasticizers is related to polyvinylchlorides or PVCs. An uPVC, or unplasticized polyvinylchloride, is used for things such as pipes. A pipe has no plasticizers in it, because it needs to remain strong and heat-resistant. Plasticized PVC is used in clothing for a flexible quality. Plasticizers are also put in some types of cling film to make the polymer more flexible.

### *Chemical Properties*

The attractive forces between polymer chains play a large part in determining a polymer's properties. Because polymer chains are so long, these interchain forces are amplified far beyond the attractions between conventional molecules. Different side groups on the polymer can lend the polymer to ionic bonding or hydrogen bonding between its own chains. These stronger forces typically result in higher tensile strength and higher crystalline melting points.

The intermolecular forces in polymers can be affected by dipoles in the monomer units. Polymers containing amide or carbonyl groups can form hydrogen bonds between adjacent chains; the partially positively charged hydrogen atoms in N-H groups of one chain are strongly attracted to the partially negatively charged oxygen atoms in C=O groups on another. These strong hydrogen bonds, for example, result in the high tensile strength and melting point of polymers containing urethane or urea linkages. Polyesters have dipole-dipole bonding between the oxygen atoms in C=O groups and the hydrogen atoms in H-C groups. Dipole bonding is not as strong as hydrogen bonding, so a polyester's melting point and strength are lower than Kevlar's (Twaron), but polyesters have greater flexibility.

Ethene, however, has no permanent dipole. The attractive forces between polyethylene chains arise from weak van der Waals forces. Molecules can be thought of as being surrounded by a cloud of negative electrons. As two polymer chains approach, their electron clouds repel one another. This has the effect of lowering the electron density on one side of a polymer chain, creating a slight positive dipole on this side. This charge is enough to attract the second polymer chain. Van der Waals forces are quite weak, however, so polyethylene can have a lower melting temperature compared to other polymers.

### *Standardized Polymer Nomenclature*

There are multiple conventions for naming polymer substances. Many commonly used polymers, such as those found in consumer products, are referred to by a common or trivial name. The trivial name is assigned based on historical precedent or popular usage rather than a standardized naming convention. Both the American Chemical Society (ACS) and IUPAC have proposed standardized naming conventions; the ACS and IUPAC conventions are similar but not identical. Examples of the differences between the various naming conventions are given in the table below:

| *Common name* | *ACS name* | *IUPAC name* |
|---|---|---|
| Poly(ethylene oxide) or PEO | Poly(oxyethylene) | Poly(oxyethene) |
| Poly(ethylene terephthalate) or PET | Poly(oxy-1,2-ethanediyloxycarbonyl-1,4-phenylenecarbonyl) | Poly(oxyetheneoxyterephthaloyl) |
| Nylon 6 | Poly[amino(1-oxo-1,6-hexanediyl)] | Poly[amino(1-oxohexan-1,6-diyl)] |

In both standardized conventions, the polymers' names are intended to reflect the monomer(s) from which they are synthesized rather than the precise nature of the repeating subunit. For example, the polymer synthesized from the simple alkene ethene is called polyethylene, retaining the *-ene* suffix even though the double bond is removed during the polymerization process:

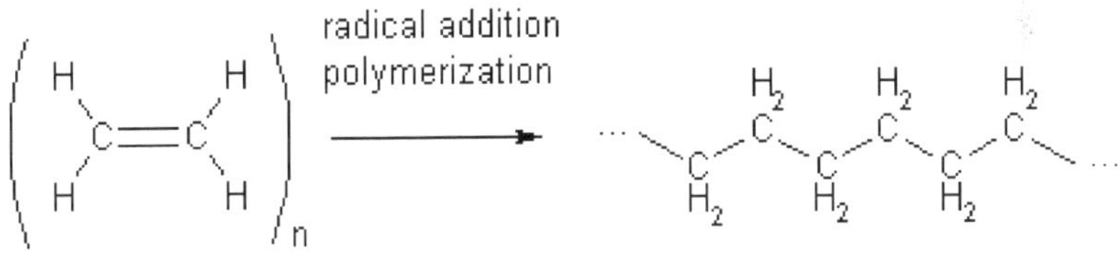


***Figure:*** *the polymerisation of ethene in to poly (ethene)*

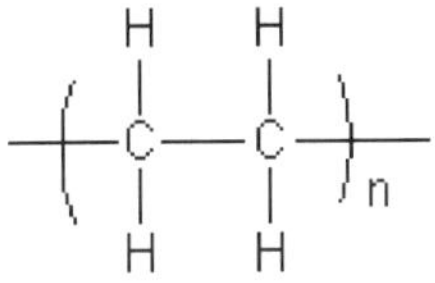

## Polymer Characterization

The characterization of a polymer requires several parameters which need to be specified. This is because a polymer actually consists of a statistical distribution of chains of varying lengths, and each chain consists of monomer residues which affect its properties.

A variety of lab techniques are used to determine the properties of polymers. Techniques such as wide angle X-ray scattering, small angle X-ray scattering, and small angle neutron scattering are used to determine the crystalline structure of polymers. Gel permeation chromatography is used to determine the number average molecular weight, weight average molecular weight, and polydispersity. FTIR, Raman and NMR can be used to determine composition. Thermal properties such as the glass transition temperature and melting point can be determined by differential scanning calorimetry and dynamic mechanical analysis. Pyrolysis followed by analysis of the fragments is one more technique for determining the possible structure of the polymer. Thermogravimetry is a useful technique to evaluate the thermal stability of the polymer. Detailed analysis of TG curves also allow us to know a bit of the phase segregation in polymers. Rheological properties are also commonly used to help determine molecular architecture (molecular weight, molecular weight distribution and branching) as well as to understand how the polymer will process, through measurements of the polymer in the melt phase. Another polymer characterization technique is Automatic Continuous Online Monitoring of Polymerization Reactions (ACOMP) which provides real-time characterization of polymerization reactions. It can be used as an analytical method in R&D, as a tool for reaction optimization at the bench and pilot plant level and, eventually, for feedback control of full-scale reactors. ACOMP measures in a model-independent fashion the evolution of average molar mass and intrinsic viscosity, monomer conversion kinetics and, in the case of copolymers, also the average composition drift and distribution. It is applicable in the areas of free radical and controlled radical homo- and copolymerization, polyelectrolyte synthesis, heterogeneous phase reactions, including emulsion polymerization, adaptation to batch and continuous reactors, and modifications of polymers.

### Polymer Degradation

Polymer degradation is a change in the properties—tensile strength, colour, shape, or molecular weight—of a polymer or polymer-based product under the influence of one or more environmental factors, such as heat, light, chemicals and, in some cases, galvanic action. It is often due to the scission of polymer chain bonds via hydrolysis, leading to a decrease in the molecular mass of the polymer.

Although such changes are frequently undesirable, in some cases, such as biodegradation and recycling, they may be intended to prevent environmental pollution. Degradation can also be useful in biomedical settings. For example, a copolymer of polylactic acid and polyglycolic acid is employed in hydrolysable stitches that slowly degrade after they are applied to a wound.

The susceptibility of a polymer to degradation depends on its structure. Epoxies and chains containing aromatic functionalities are especially susceptible to UV degradation while polyesters are susceptible to degradation by hydrolysis, while polymers containing an unsaturated backbone are especially susceptible to ozone cracking. Carbon based polymers are more susceptible to thermal degradation than inorganic polymers such as polydimethylsiloxane and are therefore not ideal for most high-temperature applications. High-temperature matrices such as bismaleimides (BMI), condensation polyimides (with an O-C-N bond), triazines (with a nitrogen (N) containing ring), and blends thereof are susceptible to polymer degradation in the form of galvanic corrosion when bare carbon fibre reinforced polymer CFRP is in contact with an active metal such as aluminium in salt water environments.

The degradation of polymers to form smaller molecules may proceed by random scission or specific scission. The degradation of polyethylene occurs by random scission—a random breakage of the bonds that hold the atoms of the polymer together. When heated above 450 °C, polyethylene degrades to form a mixture of hydrocarbons. Other polymers, such as poly (alpha-methylstyrene), undergo specific chain scission with breakage occurring only at the ends. They literally unzip or depolymerize back to the constituent monomer.

The sorting of polymer waste for recycling purposes may be facilitated by the use of the Resin identification codes developed by the Society of the Plastics Industry to identify the type of plastic.

### Product Failure

***Figure:** Ozone-induced cracking in natural rubber tubing*

In a finished product, such a change is to be prevented or delayed. Failure of safety-critical polymer components can cause serious accidents, such as fire in the case of cracked and degraded polymer fuel lines. Chlorine-induced cracking of acetal resin plumbing joints and polybutylene pipes has caused many serious floods in domestic properties, especially in the USA in the 1990s. Traces of chlorine in the water supply attacked vulnerable polymers in the plastic plumbing, a problem which occurs faster if any of the parts have been poorly extruded or injection molded. Attack of the acetal joint occurred because of faulty molding, leading to cracking along the threads of the fitting which is a serious stress concentration.

Polymer oxidation has caused accidents involving medical devices. One of the oldest known failure modes is ozone cracking caused by chain scission when ozone gas attacks susceptible elastomers, such as natural rubber and nitrile rubber. They possess double bonds in their repeat units which are cleaved during ozonolysis. Cracks in fuel lines can penetrate the bore of the tube and cause fuel leakage. If cracking occurs in the engine compartment, electric sparks can ignite the gasoline and can cause a serious fire.

Fuel lines can also be attacked by another form of degradation: hydrolysis. Nylon 6,6 is susceptible to acid hydrolysis, and in one accident, a fractured fuel line led to a spillage of diesel into the road. If diesel fuel leaks onto the road, accidents to following cars can be caused by the slippery nature of the deposit, which is like black ice.

## Varnish

***Figure:*** *Varnish on wood stairs*

Varnish is a transparent, hard, protective finish or film primarily used in wood finishing but also for other materials. Varnish is traditionally a combination of a drying oil, a resin, and a thinner or solvent. Varnish finishes are usually glossy but may be designed to produce satin or semi-gloss sheens by the addition of "flatting" agents. Varnish has little or no colour, is transparent, and has no added pigment, as opposed to paints or wood stains, which contain pigment and generally range from opaque to translucent. Varnishes are also applied over wood stains as a final step to achieve a film for gloss and protection. Some products are marketed as a combined stain and varnish.

After being applied, the film-forming substances in varnishes either harden directly, as soon as the solvent has fully evaporated, or harden after evaporation of the solvent through certain curing processes, primarily chemical reaction between oils and oxygen from the air (autoxidation) and chemical reactions between components of the varnish. Resin varnishes "dry" by evaporation of the solvent and harden almost immediately upon drying. Acrylic and waterborne varnishes "dry" upon evaporation of the water but experience an extended curing period. Oil, polyurethane, and epoxy varnishes remain liquid even after evaporation of the solvent but quickly begin to cure, undergoing successive stages from liquid or syrupy, to tacky or sticky, to dry gummy, to "dry to the touch", to hard. Environmental factors such as heat and humidity play a very large role in the drying and curing times of varnishes. In classic varnish the cure rate depends on the type of oil used and, to some extent, on the ratio of oil to resin. The drying and curing time of all varnishes may be sped up by exposure to an energy source such as sunlight, ultraviolet light, or heat. Many varnishes rely on organic oils or resins for their binder in combination with organic solvents; these are highly flammable in their liquid state. In addition, all drying oils, certain alkyds, and many single-component polyurethanes produce heat during the curing process. Therefore, oil-soaked rags and paper can smolder and ignite hours after use if they are bunched or piled together, or, for example, placed in a container where the heat cannot dissipate.

### History

Early varnishes were developed by mixing resin—pine sap, for example—with a solvent and applying them with a brush to get the golden and hardened effect one sees in today's varnishes. The ancient Egyptians were well acquainted with the art of varnishing, but its origin appears to be far east of there in India, China and Japan, where the practice of lacquer work, a species of varnish application, was known at a very early date. It has been claimed that Japan was acquainted with the art of lacquering by 500 or 600 B.C., but the majority of authorities place its first usage there to the 3rd century AD, as an art acquired from their neighbours, the Koreans. The Chinese and Indian peoples probably knew the art much earlier than the Japanese. Varnish and lacquer work are, however, generally treated in the arts as separate and distinct.

### Etymology

The word "varnish" comes from Latin *vernix* meaning odorous resin, the etymology of which comes from the Greek *Berenice,* the ancient name of modern Benghazi in Libya, where the first varnishes were used and where resins from the trees of now-vanished forests were sold. Berenice comes from the Greek words phero (to bring) + nike (victory).

### Components of Classic Varnish

***Drying oil:*** There are many different types of drying oils, including linseed oil, tung oil, and walnut oil. These contain high levels of poly unsaturated fatty acids.

### Resin

Resins that are used in varnishes include amber, kauri gum, dammar, copal, rosin (pine resin), sandarac, balsam, elemi, mastic, and others. Shellac is also a resin. In the 1900s in Canada, resins from local trees were used to finish pianos. As a result these now antique pianos are considered difficult to refinish. However, shellac can be used over the existing resins provided sufficient time is allowed for thin coats to cure. Thus the original finish can be returned to its original lustre while preserving the colour and age related crackle.

### Solvent (Traditionally Turpentine)

Traditionally, natural (organic) turpentine was used as the thinner or solvent, but has been replaced by several mineral based turpentine substitutes such as white spirit or "paint thinner", also known as "mineral spirit".

### Types

***Violin:*** Violin varnishing is a multi-step process involving some or all of the following: primer, sealer, ground, colour coats, and clear topcoat. Some systems use a drying oil varnish as described below, while others use spirit (or solvent) varnish. Touchup in repair or restoration is only done with spirit varnish.

Drying oil such as walnut oil or linseed oil may be used in combination with amber, copal, rosin or other resins. Traditionally the oil is prepared by cooking or exposure to air and sunlight, but modern "stand oil" is prepared by heating oil at high temperature without oxygen. The refined resin is sometimes available as a translucent solid and is then "run" by cooking or literally melting it in a pot over heat without solvents. The thickened oil and prepared resin are then cooked together and thinned with turpentine (away from open flame) into a brushable solution. The ingredients and processes of violin varnish are very diverse, with some highly regarded old examples showing defects (e.g. cracking, crazing) associated with incompatible varnish components.

Some violin finishing systems use vernicebianca (egg white and gum arabic) as a sealer or ground. There is also evidence that finely powdered minerals, possibly volcanic ash, were used in some grounds. Some violins made in the late 1700s, used Oxen's blood to create a very deep red colouration. Today this varnish would have faded and currently be a very warm, dark orange.

### Resin

Most resin or "gum" varnishes consist of a natural, plant- or insect-derived substance dissolved in a solvent, called *spirit varnish* or *solvent varnish*. The solvent may be alcohol, turpentine, or petroleum-based. Some resins are soluble in both alcohol and turpentine. Generally, petroleum solvents, i.e. mineral spirits or paint thinner, can substitute for turpentine. The resins include amber, dammar, copal, rosin, sandarac, elemi, benzoin, mastic, balsam, shellac, and a multitude of lacquers.

Synthetic resins such as phenolic resin may be employed as a secondary component in certain varnishes and paints. Over centuries, many recipes were developed which involved the combination of resins, oils, and other ingredients such as certain waxes. These were believed to impart special tonal qualities to musical instruments and thus were sometimes carefully guarded secrets. The interaction of different ingredients is difficult to predict or reproduce, so expert finishers were often prized professionals.

### Shellac

Shellac is a very widely used single component resin varnish that is alcohol soluble. It is not used for outdoor surfaces or where it will come into repeated contact with water such as around a sink or bathtub. The source of shellac resin is a brittle or flaky secretion of the female lac insect, *Kerrialacca*, found in the forests of Assam and

Thailand and harvested from the bark of the trees where she deposits it to provide a sticky hold on the trunk. Shellac is the basis of French polish, which for centuries has been the preferred finish for fine furniture. Specified "dewaxed" shellac has been processed to remove the waxy substances from original shellac and can be used as a primer and sanding-sealer substrate for other finishes such as polyurethanes, alkyds, oils, and acrylics.

Prepared shellac is typically available in "clear" and "amber" (or "orange") varieties, generally as "three pound cut" or three pounds dry shellac to one US gallon of alcohol. Other natural colour shades such as ruby and yellow are available from speciality pigment or woodworker's supply outlets. Dry shellac is available as refined flakes, "sticklac," "button lac," or "seedlac." "White pigmented" shellac primer paint is widely available in retail outlets, billed as a fast-drying interior primer "problem solver", in that it adheres to a variety of surfaces and seals off odors and smoke stains. Shellac clean-up may be done either with pure alcohol or with ammonia cleansers.

### Alkyd

Typically, modern commercially produced varnishes employ some form of alkyd for producing a protective film. Alkyds are chemically modified vegetable oils which operate well in a wide range of conditions and can be engineered to speed up the cure rate and thus harden faster. Better (and more expensive) exterior varnishes employ alkyds made from high performance oils and contain UV-absorbers; this improves gloss-retention and extends the lifetime of the finish. Various resins may also be combined with alkyds as part of the formula for typical "oil" varnishes that are commercially available.

### Spar Varnish

*Spar varnish* (also called *marine varnish*) was originally intended for use on ship or boatspars, to protect the timber from the effects of sea and weather. Spars bend under the load of their sails. The primary requirements were water resistance and also elasticity, so as to remain adhering as the spars flexed. Elasticity was a pre-condition for weatherproofing too, as a finish that cracked would then allow water through, even if the remaining film was impermeable. Appearance and gloss was of relatively low value. Modified tung oil and phenolic resins are often used.

When first developed, no varnishes had good UV-resistance. Even after more modern synthetic resins did become resistant, a true spar varnish maintained its elasticity above other virtues, even if this required a compromise in its UV-resistance. Spar varnishes are thus *not* necessarily the best choice for outdoor woodwork which does not need to bend in service.

Despite this, the widespread perception of "marine products" as "tough" led to domestic outdoor varnishes being branded as "Spar varnish" and sold on the virtue of their weather- and UV-resistance. These claims may be more or less realistic, depending on individual products. Only relatively recently have spar varnishes been available that can offer both effective elasticity and UV-resistance.

### Drying Oils

By definition, drying oils, such as linseed and tung oil, are not true varnishes though often in modern terms they accomplish the same thing. Drying oils cure through an exothermic reaction between the polyunsaturated portion of the oil and oxygen from the air. Originally, the term "varnish" referred to finishes that were made entirely of resin dissolved in suitable solvents, either ethanol (alcohol) or turpentine. The advantage to finishes in previous centuries was that resin varnishes had a very rapid cure rate compared to oils; in most cases they are cured practically as soon as the solvent has fully evaporated. By contrast, untreated or "raw" oils may take weeks or months to cure, depending on ambient temperature and other environmental factors. In modern terms, "boiled" or partially polymerized drying oils with added siccatives or dryers (chemical catalysts) have cure times of less than 24 hours. However, certain non-toxic by-products of the curing process are emitted from the oil film even after it is dry to the touch and over a considerable period of time. It has long been a tradition to combine drying oils with resins to obtain favourable features of both substances.

### Polyurethane

Polyurethane varnishes are typically hard, abrasion-resistant, and durable coatings. They are popular for hardwood floors but are considered by some wood finishers to be difficult or unsuitable for finishing furniture or other detailed pieces. Polyurethanes are comparable in hardness to certain alkyds but generally form a tougher film. Compared to simple oil or shellac varnishes, polyurethane varnish forms a harder, decidedly tougher and more waterproof film. However, a thick film of ordinary polyurethane may de-laminate if subjected to heat or shock, fracturing the film and leaving white patches. This tendency increases with long exposure to sunlight or when it is applied over soft woods like pine. This is also in part due to polyurethane's lesser penetration into the wood. Various priming techniques are employed to overcome this problem, including the use of certain oil varnishes, specified "dewaxed" shellac, clear penetrating epoxy sealer, or "oil-modified" polyurethane designed for the purpose.

Polyurethane varnish may also lack the "hand-rubbed" lustre of drying oils such as linseed or tung oil; in contrast, however, it is capable of a much faster and higher "build" of film, accomplishing in two coats what may require multiple applications of oil. Polyurethane may also be applied over a straight oil finish, but because of the relatively slow curing time of oils, the emission of certain chemical byproducts, and the need for exposure to oxygen from the air, care must be taken that the oils are sufficiently cured to accept the polyurethane.

Unlike drying oils and alkyds which cure, after evaporation of the solvent, upon reaction with oxygen from the air, true polyurethane coatings cure, after evaporation of the solvent, by a variety of reactions of chemicals within the original mix, or by reaction with moisture from the air. Certain polyurethane products are "hybrids" and combine different aspects of their parent components. "Oil-modified" polyurethanes, whether water-borne or solvent-borne, are currently the most widely used wood floor finishes.

Exterior use of polyurethane varnish may be problematic due to its heightened susceptibility to deterioration through ultra-violet light exposure. All clear or translucent varnishes, and indeed all film-polymer coatings (e.g. paint, stain, epoxy, synthetic plastic, etc.) are susceptible to this damage in varying degrees. Pigments in paints and stains protect against UV damage.

UV-absorbers are added to polyurethane and other varnishes (e.g. spar varnish) to work against UV damage but are decreasingly effective over the course of 2–4 years, depending on the quantity and quality of UV-absorbers added, as well as the severity and duration of sun exposure. Water exposure, humidity, temperature extremes, and other environmental factors affect all finishes. By contrast, wooden items retrieved from the Egyptian pyramids have a new and fresh appearance after 4000 years of storage. Even there, however, fungal colonies were present, and mildew and fungus are another category of entities which attack varnish. In other words, the only coat of varnish with near perfect durability is the one stored in a vacuum, in darkness, at a low and unvarying temperature. Otherwise, care and upkeep are required.

### Lacquer

The word *lacquer* refers to quick-drying, solvent-based varnishes or paints. Although their names may be similarly derived, lacquer is not the same as *shellac* and is not dissolved in alcohol. Lacquer is dissolved in lacquer thinner, which is a highly-flammable solvent typically containing butyl acetate and xylene or toluene. Lacquer is typically sprayed on, within a *spray booth* that evacuates overspray and minimizes the risk of combustion.

Outside America, the rule of thumb is that a clear wood finish formulated to be sprayed is a lacquer but if it is formulated to be brushed on then it is a varnish. Thus, by far most pieces of wooden furniture are lacquered. Lacquer may be considered different from varnish because it can be re-dissolved later by a solvent (such as the one it was dissolved in when it was applied) and does not chemically change to a solid like other varnishes.

### Acrylic

Acrylic varnishes are typically water-borne varnishes with the lowest refractive index of all finishes and high transparency. They resist yellowing. Acrylics have the advantage of water clean-up and lack of solvent fumes, but typically do not penetrate into wood as well as oils. They sometimes lack the brushability and self-levelling qualities of solvent-based varnishes. Generally they have good UV-resistance.

In the art world, varnishes offer dust-resistance and a harder surface than bare paint – they sometimes have the benefit of ultraviolet light resistors, which help protect artwork from fading in exposure to light. Acrylic varnish should be applied using an isolation coat (a permanent, protective barrier between the painting and the varnish, preferably a soft, glossy gel medium) to make varnish removal and overall conservation easier. Acrylic varnishes used for such a final removable art protection layer are typically mineral-spirit based acrylic, rather than water based.

### Two-Part

Various epoxies have been formulated as varnishes or floor finishes whereby two components are mixed directly before application. Often, the two parts are of equal volume and are referred to as "part A" and "part B". True polyurethanes are two-part systems. All two-part epoxies have a "pot-life" or "working time" during which the epoxy can be used. Usually the pot-life is a matter of a few hours but is also highly temperature dependent. Both water-borne and solvent based epoxies are used.

### *Conversion*

Used when a fast-curing, tough, hard finish is desired, such as for kitchen cabinets and office furniture. Comes in two parts - a resin and an acid catalyst. The first is a blend of an amino resin and an alkyd. The acid catalyst is added right before application in a set ratio determined by the manufacturer. Most produce minimal yellowing. There are, however, two downsides to this finish. The first is that as the finish cures, it gives off formaldehyde, which is toxic and carcinogenic. The second is that the finish can crack or craze if too many coats are applied.

## Paint

Paint is any liquid, liquefiable, or mastic composition which, after application to a substrate in a thin layer, is converted to a solid film. It is most commonly used to protect, colour or provide texture to objects.

### *History*

In 2011, South African archeologists reported finding a 100,000-year-old human-made ochre-based mixture which could have been used like paint. Cave paintings drawn with red or yellow ochre, hematite, manganese oxide, and charcoal may have been made by early *Homo sapiens* as long as 40,000 years ago.

Ancient coloured walls at Dendera, Egypt, which were exposed for years to the elements, still possess their brilliant colour, as vivid as when they were painted about 2,000 years ago. The Egyptians mixed their colours with a gummy substance, and applied them separately from each other without any blending or mixture. They appear to have used six colours: white, black, blue, red, yellow, and green. They first covered the area entirely with white, then traced the design in black, leaving out the lights of the ground colour. They used minimum for red, and generally of a dark tinge.

Pliny mentions some painted ceilings in his day in the town of Ardea, which had been done prior to the foundation of Rome. He expresses great surprise and admiration at their freshness, after the lapse of so many centuries.

Paint was made with the yolk of eggs and therefore, the substance would harden and adhere to the surface it was applied to. Pigment was made from plants, sand, and different soils. Most paints used either oil or water as a base. A still extant example of 17th-century house oil painting is Ham House in Surrey, England, where a primer was used along with several undercoats and an elaborate decorative overcoat; the pigment and oil mixture would have been ground into a paste with a mortar and pestle. The process was done by hand by the painters and exposed them to lead poisoning due to the white-lead powder.

In 1718, Marshall Smith invented a "Machine or Engine for the Grinding of Colours" in England. Although it is not known precisely how it operated, it was a device that increased the efficiency of pigment grinding dramatically. Soon, a company called Emerton and Manby was advertising exceptionally low-priced paints that had been ground with labour-saving technology:

> *One Pound of Colour ground in a Horse-Mill will paint twelve Yards of Work, whereas Colour ground any other Way, will not do half that Quantity.*

By the proper onset of the Industrial Revolution, paint was being ground in steam-powered mills and an alternative to lead-based pigments was found in a white derivative of zinc oxide. Interior house painting increasingly became the norm as the 19th century progressed, both for decorative reasons and because the paint was effective in preventing the walls rotting from damp. Linseed oil was also increasingly used as an inexpensive binder.

In 1866, Sherwin-Williams in the United States opened as a large paint-maker and invented a paint that could be used from the tin without preparation. It was not until the stimulus of World War II created a shortage of linseed oil in the supply markets that artificial resins, or alkyds, were invented. Cheap and easy to make, they also held the colour well and lasted for a long time.

## Components

***Binder, Vehicle, or Resins:*** The binder, commonly called the vehicle, is the film-forming component of paint. It is the only component that must be present. Components listed below are included optionally, depending on the desired properties of the cured film.

The binder imparts adhesion and strongly influences such properties as gloss, durability, flexibility, and toughness.

Binders include synthetic or natural resins such as alkyds, acrylics, vinyl-acrylics, vinyl acetate/ethylene (VAE), polyurethanes, polyesters, melamine resins, epoxy, or oils. Binders can be categorized according to the mechanisms for drying or curing. Although drying may refer to evaporation of the solvent or thinner, it usually refers to oxidative cross-linking of the binders and is indistinguishable from curing. Some paints form by solvent evaporation only, but most rely on cross-linking processes.

Paints that dry by solvent evaporation and contain the solid binder dissolved in a solvent are known as lacquers. A solid film forms when the solvent evaporates, and because the film can re-dissolve in solvent, lacquers are unsuitable for applications where chemical resistance is important. Classic nitrocellulose lacquers fall into this category, as do non-grain raising stains composed of dyes dissolved in solvent and more modern acrylic-based coatings such as 5-ball Krylon aerosol. Performance varies by formulation, but lacquers generally tend to have better UV resistance and lower corrosion resistance than comparable systems that cure by polymerization or coalescence.

The paint type known as Emulsion in the UK and Latex in the USA is a water-borne dispersion of sub-micrometre polymer particles. These terms in their respective countries cover all paints that use synthetic polymers such as acrylic, vinyl acrylic (PVA), styrene acrylic, etc. as binders. The term "latex" in the context of paint in the USA simply means an aqueous dispersion; latex rubber from the rubber tree is not an ingredient. These dispersions are prepared by emulsion polymerization. Such paints cure by a process called coalescence where first the water, and then the trace, or coalescing, solvent, evaporate and draw together and soften the binder particles and fuse them together into irreversibly bound networked structures, so that the paint will not redissolve in the solvent/water that originally carried it. The residual surfactants in paint as well as hydrolytic effects with some polymers cause the paint to remain susceptible to softening and, over time, degradation by water. The general term of latex paint is usually used in the USA, while the term emulsion paint is used for the same products in the UK and the term latex paint is not used at all. Paints that cure by oxidative crosslinking are generally single package coatings. When applied, the exposure to oxygen in the air starts a process that crosslinks and polymerizes the binder component. Classic alkyd enamels would fall into this category. Oxidative cure coatings are catalyzed by metal complex driers such as cobalt naphthenate.

Paints that cure by polymerization are generally one or two package coatings that polymerize by way of a chemical reaction and which cure into a crosslinked film. Depending on composition they may need to dry first, by evaporation of solvent. Classic two package epoxies or polyurethanes would fall into this category.

There are paints called plastisols/organosols, which are made by blending PVC granules with a plasticiser. These are stoved and the mix coalesces. Other films are formed by cooling of the binder. For example, encaustic or wax paints are liquid when warm, and harden upon cooling. In many cases, they will resoften or liquify if reheated.

Recent environmental requirements restrict the use of volatile organic compounds (VOCs), and alternative means of curing have been developed, particularly for industrial purposes. In UV curing paints, the solvent is evaporated first, and hardening is then initiated by ultraviolet light. In powder coatings there is little or no solvent, and flow and cure are produced by heating of the substrate after electrostatic application of the dry powder.

### Diluent or Solvent

The main purposes of the diluent are to dissolve the polymer and adjust the viscosity of the paint. It is volatile and does not become part of the paint film. It also controls flow and application properties, and in some cases can affect the stability of the paint while in liquid state. Its main function is as the carrier for the non volatile components. To spread heavier oils (for example, linseed) as in oil-based interior house paint, a thinner oil is required. These volatile substances impart their properties temporarily—once the solvent has evaporated, the remaining paint is fixed to the surface.

This component is optional: some paints have no diluent. Water is the main diluent for water-borne paints, even the co-solvent types.

Solvent-borne, also called oil-based, paints can have various combinations of organic solvents as the diluent, including aliphatics, aromatics, alcohols, ketones and white spirit. Specific examples are organic solvents such as petroleum distillate, esters, glycol ethers, and the like. Sometimes volatile low-molecular weight synthetic resins also serve as diluents.

### Pigment or Filler

Pigments are granular solids incorporated in the paint to contribute colour. Fillers are granular solids incorporate to impart toughness, texture, give the paint special properties, or to reduce the cost of the paint. Alternatively, some paints contain dyes instead of or in combination with pigments.

Pigments can be classified as either natural or synthetic. Natural pigments include various clays, calcium carbonate, mica, silicas, and talcs. Synthetics would include engineered molecules, calcined clays, blanc fixe, precipitated calcium carbonate, and synthetic pyrogenic silicas. Hiding pigments, in making paint opaque, also protect the substrate from the harmful effects of ultraviolet light. Hiding pigments include titanium dioxide, phthalo blue, red iron oxide, and many others.

Fillers are a special type of pigment that serve to thicken the film, support its structure and increase the volume of the paint. Fillers are usually cheap and inert materials, such as diatomaceous earth, talc, lime, barytes, clay, etc. Floor paints that will be subjected to abrasion may contain fine quartz sand as a filler. Not all paints include fillers. On the other hand, some paints contain large proportions of pigment/filler and binder.

Some pigments are toxic, such as the lead pigments that are used in lead paint. Paint manufacturers began replacing white lead pigments with titanium white (titanium dioxide), before lead was banned in paint for residential use in 1978 by the US Consumer Product Safety Commission. The titanium dioxide used in most paints today is often coated with silica/alumina/zirconium for various reasons, such as better exterior durability, or better hiding performance (opacity) promoted by more optimal spacing within the paint film.

### Additives

Besides the three main categories of ingredients, paint can have a wide variety of miscellaneous additives, which are usually added in small amounts, yet provide a significant effect on the product. Some examples include additives to modify surface tension, improve flow properties, improve the finished appearance, increase wet edge, improve pigment stability, impart antifreeze properties, control foaming, control skinning, etc. Other types of additives include catalysts, thickeners, stabilizers, emulsifiers, texturizers, adhesion promoters, UV stabilizers, flatteners (de-glossing agents), biocides to fight bacterial growth, and the like.

Additives normally do not significantly alter the percentages of individual components in a formulation.

### Colour Changing Paint

Various technologies exist for making paints that change colour. Thermochromic paints and coatings contain materials that change conformation when heat is applied, and so they change colour. Liquid crystals have been used in such paints, such as in the thermometre strips and tapes used in aquaria. Photochromic paints and coatings contain dyes that change conformation when the film is exposed to UV light, and so they change colour. These materials are used to make eyeglasses.

Colour changing paints can also be made by adding halochrome compounds or other organic pigments. One patent cites use of these indicators for wall coating applications for light coloured paints. When the paint is wet it is pink in colour but upon drying it regains its original white colour. As cited in patent, this property of the paint enabled two or multiple coats to be applied on a wall properly and evenly. The previous coat/s having dried would be white whereas the new wet coat would be distinctly pink. Ashland Inc. introduced foundry refractory coatings with similar principle in 2005 for use in foundries.

Electrochromic paints change colour in response to an applied electric current. Car manufacturer Nissan has been reportedly working on an electrochromic paint, based on particles of paramagneticiron oxide. When subjected to an electromagnetic field the paramagnetic particles change spacing, modifying their colour and reflective properties. The electromagnetic field would be formed using the conductive metal ofthe car body. Electrochromic paints can be applied to plastic substrates as well, using a different coating chemistry. The technology involves using special dyes that change conformation when an electric current is applied across the film itself. Recently, this new technology has been used to achieve glare protection at the touch of a button in passenger airplane windows.

### Art

***Figure:*** *Watercolours as applied with a brush*

Since the time of the Renaissance, siccative(drying) oil paints, primarily linseed oil, have been the most commonly used kind of paints in fine art applications; oil paint is still common today. However, in the 20th century, water-based paints, including watercolours and acrylic paints, became very popular with the development of acrylic and other latex paints. Milk paints (also called casein), where the medium is derived from the natural emulsion that is milk, were popular in the 19th century and are still available today. Egg tempera (where the medium is an emulsion of raw egg yolk mixed with oil) is still in use as well, as are encaustic wax-based paints. Gouache is a variety of opaque watercolour which was also used in the Middle Ages and Renaissance for manuscript illuminations. The pigment was often made from ground semiprecious stones such as lapis lazuli and the binder made from either gum arabic or egg white. Gouache, also known as 'designer colour' or 'body colour' is commercially available today.

Poster paint has been used primarily in the creation of student works, or by children.

### *Application*

Paint can be applied as a solid, a gaseous suspension (aerosol) or a liquid. Techniques vary depending on the practical or artistic results desired. As a solid (usually used in industrial and automotive applications), the paint is applied as a very fine powder, then baked at high temperature. This melts the powder and causes it to adhere to the surface. The reasons for doing this involve the chemistries of the paint, the surface itself, and perhaps even the chemistry of the substrate (the object being painted). This is called "powder coating" an object.

As a gas or as a gaseous suspension, the paint is suspended in solid or liquid form in a gas that is sprayed on an object. The paint sticks to the object. This is called "spray painting" an object. The reasons for doing this include:

- The application mechanism is air and thus no solid object touches the object being painted;
- The distribution of the paint is uniform, so there are no sharp lines;
- It is possible to deliver very small amounts of paint;
- A chemical (typically a solvent) can be sprayed along with the paint to dissolve together both the delivered paint and the chemicals on the surface of the object being painted;

- Some chemical reactions in paint involve the orientation of the paint molecules.

In the liquid application, paint can be applied by direct application using brushes, paint rollers, blades, other instruments, or body parts such as fingers and thumbs.

Rollers generally have a handle that allows for different lengths of poles to be attached, allowing painting at different heights. Generally, roller application requires two coats for even colour. A roller with a thicker nap is used to apply paint on uneven surfaces. Edges are often finished with an angled brush.

- Using the finish flat one would most likely use a 1/2" nap roller
- Using the finish eggshell one would most likely use a 3/8" nap roller
- Using the finish satin or pearl one would most likely use a 3/8" nap roller
- Using the finish semi-gloss or gloss one would most likely use a 3/16" nap roller

After liquid paint is applied, there is an interval during which it can be blended with additional painted regions (at the "wet edge") called "open time." The open time of an oil or alkyd-based emulsion paint can be extended by adding white spirit, similar glycols such as Dowanol (propylene glycol ether) or open time prolongers. This can also facilitate the mixing of different wet paint layers for aesthetic effect. Latex and acrylic emulsions require the use of drying retardants suitable for water-based coatings.

Paint application by spray is the most popular method in industry. In this, paint is atomized by the force of compressed air or by the action of high pressure compression of the paint itself, and the paint is turned into small droplets which travel to the article which is to be painted. Alternate methods are airless spray, hot spray, hot airless spray, and any of these with an electrostatic spray included. There are numerous electrostatic methods available.

Dipping used to be the norm for objects such as filing cabinets, but this has been replaced by high speed air turbine driven bells with electrostatic spray. Car bodies are primed using cathodicelephoretic primer, which is applied by charging the body depositing a layer of primer. The unchanged residue is rinsed off and the primer stoved.

Many paints tend to separate when stored, the heavier components settling to the bottom, and require mixing before use. Some paint outlets have machines for mixing the paint by shaking the can vigorously for a few minutes.

The opacity and the film thickness of paint may be measured using a drawdown card. Water-based paints tend to be the easiest to clean up after use; the brushes and rollers can be cleaned with soap and water.

Proper disposal of left over paint is a challenge. Sometimes it can be recycled: Old paint may be usable for a primer coat or an intermediate coat, and paints of similar chemistry can be mixed to make a larger amount of a uniform colour.

To dispose of paint it can be dried and disposed of in the domestic waste stream, provided that it contains no prohibited substances. Disposal of liquid paint usually requires special handling and should be treated as hazardous waste, and disposed of according to local regulations.

### *Product Variants*

- Primer is a preparatory coating put on materials before painting. Priming ensures better adhesion of paint to the surface, increases paint durability, and provides additional protection for the material being painted. It can also be used to block and seal stains, or to hide a colour that is to be painted over.
- Emulsion paint is a water-based paint used for painting interior or exterior surfaces.
- Flat Finish-This finish is generally used on ceilings or walls that are in bad shape. This finish is more forgiving in terms of hiding imperfections in your walls and covers greatly. However this finish provides no washability if you were to get a stain on the wall.
- Matte Finish-This finish is very similar if not the same to the flat finish, but comes with some washability and good coverage.
- Eggshell Finish-This finish has some sheen to it, the equivalence to a shell on an egg. This finish provides great washability, but shows some imperfections on walls. This finish is ideal for bathrooms because it offers not only washability, but has enough shine so water will not stick to it and make it peel.
- Pearl (Satin) Finish-This is very durable in terms of washability and resistance to moisture. It will offer the user the full protection on their walls from dirt, moisture, stains, etc. This

finish is ideal for bathrooms, furniture, and kitchens, but will show even more imperfections on walls because of its strong shine (this finish is shinier then eggshell).

- Semi-Gloss Finish-This finish is typically used on trim to add detail, elegance, and to show off the wood work. It is used on doors and furniture as well. This is a very high shine, but provides the most protection from moisture and stains on your walls. Obviously this finish will show the imperfections on the wall tremendously. This finish is generally used in schools and factories where washability and durability are the only considerations.
- Varnish and shellac provide a protective coating without changing the colour. They are paints without pigment.
- Wood stain is a type of paint that is very "thin," that is, low in viscosity, and formulated so that the pigment penetrates the surface rather than remaining in a film on the surface. Stain is predominantly pigment or dye and solvent with little binder, designed primarily to add colour without providing a surface coating.
- Lacquer is usually a fast-drying solvent-based paint or varnish that produces an especially hard, durable finish.
- An enamel paint is a paint that dries to an especially hard, usually glossy, finish. Enamel paints sometimes contain glass powder or tiny metal flake fragments instead of the colour pigments found in standard oil-based paints. Enamel paint is sometimes mixed with varnish or urethane to increase shine as well as assist its hardening process.
- A glaze is an additive used with paint to slow drying time and increase translucency, as in faux painting and Art Painting.
- A roof coating is a fluid applied membrane which has elastic properties that allows it to stretch and return to their original shape without damage. It provides UV protection to polyurethane foam and is widely used as part of a roof restoration system.
- Fingerpaint is a kind of paint intended to be applied with the fingers; it typically comes in pots and is normally used by small children.
- Inks are similar to paints, except they are typically made using finely ground pigments or dyes, and are designed so as not to leave a thick film of binder.

- Anti-graffiti coatings are used to defeat the marking of surfaces by graffiti vandals. There are two categories, sacrificial and non-bonding. Sacrificial coatings are clear coatings that allow the removal of graffiti, usually by pressure washing the surface with high-pressure water, removing the graffiti, and the coating (hence, sacrificed). They must be re-applied afterward for continued protection.

  This is most commonly used on natural-looking masonry surfaces, such as statuary and marble walls, and on rougher surfaces that are difficult to clean. Non-bonding coatings are clear, high-performance coatings, usually catalyzed polyurethanes, that allow the graffiti very little to bond to. After the graffiti is discovered, it can be removed with the use of a solvent wash, without damaging the underlying substrate or protective coating. These work best when used on smoother surfaces, and especially over other painted surfaces, including murals.
- Anti-climb paint is a non-drying paint that appears normal while being extremely slippery. It is usually used on drainpipes and ledges to deter burglars and vandals from climbing them, and is found in many public places. When a person attempts to climb objects coated with the paint, it rubs off onto the climber, as well as making it hard for them to climb.
- Anti-fouling paint, or bottom paint, prevents barnacles and other marine organisms from adhering to the hulls of ships.
- Insulative paint, or insulating paint, reduces the rate of thermal transfer through a surface to which is applied. It consists of any paint with an additive of microspheres that are hollow (vacuum-filled).

### *Failure of a Paint*

The main reasons of paint failure after application on surface are the applicator and improper treatment of surface.

Application Defects can be attributed to:

***Dilution:*** This usually occurs when the dilution of the paint is not done as per manufacturers recommendation. There can be a case of over dilution and under dilution, as well as dilution with the incorrect diluent.

***Contamination:*** Foreign contaminants added without the manufacturers consent which results in various film defects.

***Peeling/Blistering:*** Most commonly due to improper surface treatment before application and inherent moisture/dampness being present in the substrate.

***Chalking:*** Chalking is the progressive powdering of the paint film on the painted surface. The primary reason for the problem is polymer degradation of the paint matrix due to exposure of UV radiation in sunshine and condensation from dew. The degree of chalking varies as epoxies react quickly while acrylics and polyurethanes can remain unchanged for long periods. The degree of chalking can be assessed according to International Standard ISO 4628 Part 6 or 7 or American Society of Testing and Materials(ASTM) Method D4214 (Standard Test Methods for Evaluating the Degree of Chalking of Exterior Paint Films).

***Cracking:*** Cracking of paint film is due to the unequal expansion or contraction of paint coats. It usually happens when the coats of the paint are not allowed to cure/dry completely before the next coat is applied. The degree of cracking can be assessed according to International Standard ISO 4628 Part 4 or ASTM Method D661 (Standard Test Method for Evaluating Degree of Cracking of Exterior Paints).

***Erosion:*** Erosion is very quick chalking. It occurs due to external agents like air,water etc. It can be evaluated using ASTM Method ASTM D662 (Standard Test Method for Evaluating Degree of Erosion of Exterior Paints).

***Blistering:*** Blistering is due to improper surface exposure of paint to strong sunshine. The degree of blistering can be assessed according to ISO 4628 Part 2 or ASTM Method D714 (Standard Test Method for Evaluating Degree of Blistering of Paints).

## Dangers

Volatile organic compounds (VOCs) in paint are considered harmful to the environment and especially for people who work with them on a regular basis. Exposure to VOCs has been related to organic solvent syndrome, although this relation has been somewhat controversial.

In the US, environmental regulations, consumer demand, and advances in technology led to the development of low-VOC and zero-VOC paints and finishes. These new paints are widely available and meet or exceed the old high-VOC products in performance and cost-effectiveness while having significantly less impact on human and environmental health.

# 7

# Building Insulation

Building insulation refers broadly to any object in a building used as insulation for any purpose. While the majority of insulation in buildings is for thermal purposes, the term also applies to acoustic insulation, fire insulation, and impact insulation (e.g. for vibrations caused by industrial applications). Often an insulation material will be chosen for its ability to perform several of these functions at once.

## Thermal Insulation

Thermal insulation in buildings is an important factor to achieving thermal comfort for its occupants. Insulation reduces unwanted heat loss or gain and can decrease the energy demands of heating and cooling systems. It does not necessarily deal with issues of adequate ventilation and may or may not affect the level of sound insulation. In a narrow sense insulation can just refer to the insulation materials employed to slow heat loss, such as: cellulose, glass wool, rock wool, polystyrene, urethane foam, vermiculite, perlite, wood fibre, plant fibre (cannabis, flax, cotton, cork, etc.), plant straw, animal fibre (sheep's wool), cement, and earth or soil, but it can also involve a range of designs and techniques to address the main modes of heat transfer - conduction, radiation and convection materials.

The effectiveness of insulation is commonly evaluated by its R-value, of which there are two - metric (SI) and US customary, the former being 0.176 times the latter. For attics, it is recommended that it should be at least R-38 (US customary, R-6.7 metric). However, an R-value does not take into account the quality of construction or local environmental factors for each building. Construction quality issues include inadequate vapor barriers, and problems with draft-proofing.

In addition, the properties and density of the insulation material itself is critical.

### Planning

How much insulation a house should have depends on building design, climate, energy costs, budget, and personal preference. Regional climates make for different requirements. Building codes specify only the bare minimum; insulating beyond what the code requires is often recommended.

The insulation strategy of a building needs to be based on a careful consideration of the mode of energy transfer and the direction and intensity in which it moves. This may alter throughout the day and from season to season. It is important to choose an appropriate design, the correct combination of materials and building techniques to suit the particular situation. To determine whether you should add insulation, you first need to find out how much insulation you already have in your home and where. A qualified home energy auditor will include an insulation check as a routine part of a whole-house energy audit.

### In the USA

An initial estimate of insulation needs in the United States can be determined by the US Department of Energy's ZIP codeinsulation calculator.

### Russia

In Russia the luxury of cheap gas has led to poorly insulated, overheated and inefficient consumers of energy. The Russian Centre for Energy Efficiency found that Russian buildings are either over- or under-heated, often consume up to 50 percent more heat and hot water than needed. Fifty-three percent of all carbon dioxide ($CO_2$) emissions in Russia are produced through heating and generating electricity for buildings. However, Green House Gas (GHG) emissions from the Soviet Bloc are still below their 1990 levels).

### Climate

***Cold climates:*** In cold conditions, the main aim is to reduce heat flow out of the building. The components of the building envelope - windows, doors, roofs, walls, and air infiltration barriers - are all important sources of heat loss; in an otherwise well insulated home, windows will then become an important source of heat transfer. The resistance to conducted heat loss for standard glazing corresponds to an R-value of about 0.17W/m²/K° (compared to 2-4W/m²/K° for glasswoolbatts). Losses can be reduced by good weatherisation, bulk insulation, and minimising the amount of non-insulative (particularly non-solar facing) glazing. Indoor thermal radiation can also be a disadvantage with spectrally selective (low-e, low-emissivity) glazing. Some insulated glazing systems can double to triple R values.

***Figure:*** *Cross-section of home insulation.*

### Hot Climates

In hot conditions, the greatest source of heat energy is solar radiation. This can enter buildings directly through windows or it can heat the building shell to a higher temperature than the ambient, increasing the heat transfer through the building envelope. The Solar Heat Gain Co-efficient (SGHC) (a measure of solar heat transmittance) of standard single glazing can be around 78-85%. Solar gain can be reduced by adequate shading from the sun, light coloured roofing, spectrally selective (heat-reflective) paints and coatings and various types of insulation for the rest of the envelope. Specially coated glazing can reduce SHGC to around 10%. Radiant barriers are highly effective for attic spaces in hot climates. In this application, they are much more effective in hot climates than cold climates. For downward heat flow, convection is weak and radiation dominates heat transfer across an air space. Radiant barriers must face an adequate air-gap to be effective.

If refrigerative air-conditioning is employed in a hot, humid climate, then it is particularly important to seal the building envelope. Dehumidification of humid air infiltration can waste significant energy. On the other hand, some building designs are based on effective cross-ventilation instead of refrigerative air-conditioning to provide convective cooling from prevailing breezes.

### *Orientation - Passive Solar Design*

Optimal placement of building elements (e.g. windows, doors, heaters) can play a significant role in insulation by considering the impact of solar radiation on the building and the prevailing breezes. Reflective laminates can help reduce passive solar heat in pole barns, garages and metal buildings.

### *Building Envelope*

The thermal envelope defines the conditioned or living space in a house. The attic or basement may or may not be included in this area. Reducing airflow from inside to outside can help to reduce convective heat transfer significantly. Ensuring low convective heat transfer also requires attention to building construction (weatherization) and the correct installation of insulative materials.

The less natural airflow into a building, the more mechanical ventilation will be required to support human comfort. High humidity can be a significant issue associated with lack of airflow, causing condensation, rotting construction materials, and encouraging microbial growth such as mould and bacteria. Moisture can also drastically reduce the effectiveness of insulation by creating a thermal bridge. Air exchange systems can be actively or passively incorporated to address these problems.

Some of the unwanted continuous air exchange with the outside air, which is an artifact of the inherent hysterisis of thermostat control systems, is inevitable. These thermostats typically require the indoor air temperature to cycle 1 to 2 degrees Fahrenheit (about 1 Celsius) as the heating system cycles on and off. Each of these cycles results in a 0.2% to 0.4% volume exchange with the outside air as conditioned envelope air expands and contracts.

Thus a 3000 square foot home with 9 foot ceilings having a volume of 27,000 cubic feet and a heating/cooling system cycling every ten minutes will exchange 7700 to 15400 cubic feet in a 24 hr period. This incoming, unconditioned air places an additional load on heating and cooling systems and results in additional costs to energy consumers. Interestingly, the cycling frequency of the heating/cooling system is at maximum when the heating/cooling load equals half the heating/cooling system capacity.

### *Thermal Bridge*

Thermal bridges are points in the building envelope that allow heat conduction to occur. Since heat flows through the path of least resistance, thermal bridges can contribute to poor energy performance. A thermal bridge is created when materials create a continuous path across a temperature difference, in which the heat flow is not interrupted by thermal insulation. Common building materials that are poor insulators include glass and metal.

A building design may have limited capacity for insulation in some areas of the structure. A common construction design is based on stud walls, in which thermal bridges are common in wood or steel studs and joists, which are typically fastened with metal. Notable areas that most commonly lack sufficient insulation are the corners of buildings, and areas where insulation has been removed or displaced to make room for system infrastructure, such as electrical boxes (outlets and light switches), plumbing, fire alarm equipment, etc.

Thermal bridges can also be created by uncoordinated construction, for example by closing off parts of external walls before they are fully insulated. The existence of inaccessible voids within the wall cavity which are devoid of insulation can be a source of thermal bridging.

Some forms of insulation transfer heat more readily when wet, and can therefore also form a thermal bridge in this state.

The heat conduction can be minimized by any of the following: reducing the cross sectional area of the bridges, increasing the bridge length, or decreasing the number of thermal bridges.

One method of reducing thermal bridge effects is the installation of an insulation board (e.g. foam board EPS XPS, wood fibre board, etc.) over the exterior outside wall. Another method is using insulated lumber framing for a thermal break inside the wall.

### Materials

There are essentially two types of building insulation - Bulk Insulation and Reflective Insulation. Most buildings use a combination of both types to make up a total building insulation system. The type of insulation used is matched to create maximum resistance to each of the three forms of building heat transfer - Conduction, Convection, and Radiation.

### Conductive and Convective Insulators ('Bulk Insulation')

Bulk insulators block conductive heat transfer and convective flow either into or out of a building. The denser a material is, the better it will conduct heat. Because air has such low density, air is a very poor conductor and therefore makes a good insulator. Insulation to resist conductive heat transfer uses air spaces between fibres, inside foam or plastic bubbles and in building cavities like the attic. This is beneficial in an actively cooled or heated building, but can be a liability in a passively cooled building; adequate provisions for cooling by ventilation or radiation are needed.

### Radiant Heat Barriers

Radiant barriers work in conjunction with an air space to reduce radiant heat transfer across the air space. Radiant or reflective insulation reflects heat instead of either absorbing it or letting it pass through. Radiant barriers are often seen used in reducing downward heat flow, because upward heat flow tends to be dominated by convection. This means that for attics, ceilings, and roofs, they are most effective in hot climates. They also have a role in reducing heat losses in cool climates. However, much greater insulation can be achieved through the addition of bulk insulators.

Some radiant barriers are spectrally selective and will preferentially reduce the flow of infra-red radiation in comparison to other wavelengths. For instance low-emissivity (low-e) windows will transmit light and short-wave infra-red energy into a building but reflect back the long-wave infra-red radiation generated by interior furnishings. Similarly, special heat-reflective paints are able to reflect more heat than visible light, or vice-versa.

Thermal emissivity values probably best reflect the effectiveness of radiant barriers. Some manufacturers quote an 'equivalent' R-value for these products but these figures can be difficult to interpret, or even misleading, since R-value testing measures total heat loss in a laboratory setting and does not control the type of heat loss responsible for the net result (radiation, conduction, convection). A film of dirt or moisture can alter the emissivity and hence the performance of radiant barriers.

### Installation of Insulation

Insulating buildings during construction is much easier than retrofitting, as generally the insulation is hidden, and parts of the building need to be deconstructed to reach them.

## Wood Stain

A wood stain consists of a colourant suspended or dissolved in an agent or solvent. The suspension agent can be water, alcohol, petroleum distillate, or the actual finishing agent (shellac, lacquer, varnish, polyurethane, etc.). Coloured or 'stained' finishes, like polyurethane, do not penetrate the pores of the wood to any significant degree and will disappear when the finish itself deteriorates or is removed intentionally.

Pigments and dyes are largely used as colourants. The difference between the two is in the size of the particles. Dyes are microscopic crystals that dissolve in the vehicle and pigments are suspended in the vehicle and are much larger. Dyes will colour very fine grained wood, like cherry or maple, which pigments will not. Those fine-grained woods have pores too small for pigments to attach themselves to. Pigments contain a binder to help attach themselves to the wood.

The type of stain will either accentuate or obscure the wood grain. Most commercial stains contain both dye and pigment and the degree to which they stain the appropriate wood is mostly dependent on the length of time they are left on the wood. Pigments, regardless of the suspension agent, will not give much colour to very dense woods but will deeply colour woods with large pores (*e.g.* pine). Dyes are translucent and pigments are opaque.

Gel stains are more akin to paint and have little penetrating ability.

### Composition

Stain is composed of the same three primary ingredients as paint (pigment, solvent, and binder) but is predominantly pigment (or dye) and solvent with little binder. Much like the dyeing or staining of fabric, wood stain is designed to add colour to the substrate of wood and other materials while leaving the substrate mostly visible. Transparent varnishes or surface films are applied afterwards. In principle, stains do not provide a surface coating or film. However, because the binders are from the same class of film-forming binders that are used in paints and varnishes, some build-up of film occurs.

### Comparison with Paint and Varnish

The initial application of any paint or varnish is similarly absorbed into the substrate, but because stains contain lower amounts of binder, the binder from a stain resides mainly below the surface while the pigment remains near the top or at the surface. Stains that employ metallic pigments such as iron oxides usually are more opaque; first because metallic pigments are opaque by nature, but also because the particles of which they consist are much larger than organic pigments and therefore do not penetrate as well. Most wood stains for interior uses (e.g. floors and furniture) require further application of varnish or finish for protection and gloss. Stains are differentiated from varnishes in that the latter has no added colour or pigment and is designed to form a surface film. Some products are marketed as a combination of stain and varnish.

### Siding Stain

Siding stain is one variety of wood stain with very high viscosity (others can be quite thin). Effectively, siding stains are paints that do not cover as well and do not form a hard film. They are designed to penetrate better and contain binders that are softer and more flexible, allowing them to last longer than harder, more brittle paints. Siding stain protects against solar radiation especially UV radiation, water, fungus including mildew, and insects. Different siding stains are distinguished by the appearance they impart to wood. Certain solvent-based or oil-based siding stains contain small amounts of paraffin wax, which cannot be painted over, although re-staining is still possible.

### Absorption

Applying stains can be very easy or very difficult depending on the type of substrate, type of stain, and the ambient conditions. Fresh, "green" lumber accepts stain poorly, while aged wood absorbs stains relatively well. Porosity of wood can vary greatly, even within the same piece of wood. End grain and bias-cut grain are far more absorbent, thus will accept more pigment and will darken considerably in those areas. The hard ring may absorb differently from the soft ring. The characteristic medullary rays in oak will absorb much less and remain mostly blonde. Woods that have been heavily subjected to paint strippers or washed down with detergents or solvents will have an increased open grain and accept substantially more stain than normal. Woods from different species of trees can have huge variations in how well they take stain. Different wood species stain differently—the overall colour and shade is a result of a combination of the stain and properties of the wood. For example, although medium-to-dark stains tend to look blotchy on maple, they get deeper and more glowing on cherry, with a more consistent colouration. Stains that are fast drying will be difficult to apply in hot weather or in direct sunlight. Stains that are slow-drying will be difficult to work with in damp and cold conditions due to a greatly lengthened evaporation and curing period. New lumber, such as pine, can have waxlike sealants put on at the mill that will prevent proper staining; stripping or sanding the surface may be required. White stains composed of metal oxides, namely titanium dioxide and zinc oxide do not penetrate well and remain on the surface. In such cases, wear easily reveals unstained wood. They are also fairly opaque.

### *Preparation*

Typically, a thorough final sanding of the wood with one suitable grit will "even out" the absorption of the stain. White stain on a bare softwood or oak floor might require a final 'prep' sanding by hand with an orbital/vibrating sander with 80 or 100 grit, whereas certain hardwoods might be orbitally or hand sanded with 220 grit and higher for a darkish organic stain on furniture. Though it is not as durable, "garnet" is by far the preferred sandpaper for hand-sanding bare wood; first because it is a sharper and faster cutting grit, and second because it does not impart colour from the grit, as aluminium oxide or silicon carbide can do. Other methods include "rubbing" with rotten-stone etc.

In certain cases you need to clean the wood or remove existing stains prior to staining the wood with a commercial stain in order to avoid damaging the wood. This can be the case for both unfinished and finished wood.

### *Cleaning Stained Wood*

There are special consideration when cleaning stained wood. One of the most common stains is water stains on stained wood. Techniques to remove the water stains have been documented that use a hot iron to remove the water stain. Special caution should be used when trying to remove stains from a stained wood to avoid damaging the original stain. This is especially important with antiques.

## Glass Brick

Glass brick, also known as glass block, is an architectural element made from glass. Glass bricks provide visual obscuration while admitting light. The glass block was originally developed in the early 1900s to provide natural light in manufacturing plants. Glass bricks are produced for both wall and floor applications. Glass blocks for use in floors are normally manufactured as a single solid piece, or as a hollow glass block with thicker side walls than the standard wall blocks. These blocks are normally cast into a reinforced concrete gridwork or set into a metal frame, allowing multiple units to be combined to span over openings in basements and roofs. Glass wall blocks should not be used in flooring applications. Hollow glass wall blocks are manufactured as two separate hemispheres and, whilst the glass is still molten, the two hemispheres are pressed together and annealed. The resulting glass blocks will have a partial vacuum at the hollow centre.

Specialist glass blocks are produced for various applications including:

Bullet and vandal resistance: these blocks are generally solid glass or have very thick side walls similar to pavement blocks.

Fire resistance: fire resistance of varying degrees can be achieved by several methods. Standard production hollow wall block will offer little fire resistance; however, resistance is improved by utilising specially produced hollow blocks with thicker sidewalls, or the inclusion of a special layer of fire resisting material between the two hemispheres of the block during manufacture. Some manufacturers of glass blocks have developed a method of bonding two glass blocks together with adhesive, producing blocks of up to 160 mm (6½") thick with enhanced fire resistance. It is important that the block manufacturer's recommendations are followed with regards to the installation of fire resisting glass block walls, as without special construction techniques, the wall will not achieve the desired fire resistance.

A recent innovation in the manufacture of glass blocks is the inclusion of argon gas within the hollow centre of glass wall blocks. This advancement in production technique has resulted in a glass block which is able to offer significantly improved thermal insulation properties.

Glass blocks in Europe are manufactured in accordance with the European Standard EN1052-2. The International Standard is ISO TC 160/SG1. The Standards allow for variation in sizes and production irregularity. Blocks fall within three classifications; Class1, Class 2 and Class 3 with Class 1 being the highest and best rating with a maximum permissible deviation from designed size and rectangularity of 1 mm. Some hollow glass wall blocks are available in coloured variants. These coloured variants fall into two categories; those that are manufactured with coloured glass which are UV stable and can be used in the same locations as standard clear glass blocks.

The other method by which coloured glass blocks are achieved is to inject a coloured material, dye or transparent paint into the hollow centre of the blocks to form a permanent coating. This method of producing coloured blocks enables vibrant colours to be achieved which are not possible with coloured glass. The downside of this production method is that the coloured coating may not be UV stable and can fade in bright sunshine over time and may therefore, not be suitable for all locations.

Glass wall blocks are fixed together to form complete walls by several methods – the most common method of construction is to bed the blocks together in a Portland cement-based mortar with reinforcing rods of steel placed within the mortar as recommended by the project architect or block manufacturer. Other methods of construction include several proprietary systems whereby the mortar is replaced by timber or PVC extrusions.

## Structural Steel

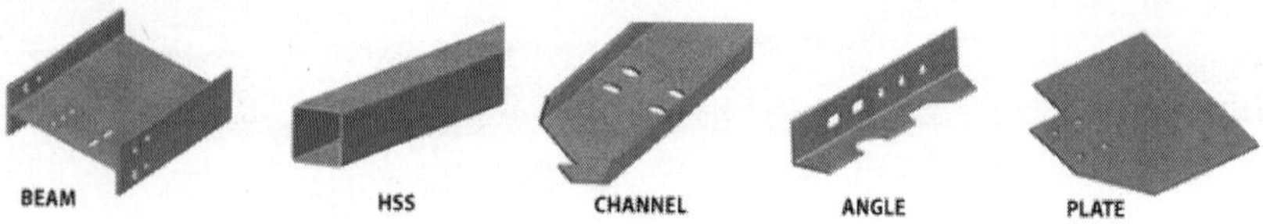


***Figure:*** *Various structural steel shapes*

Structural steel is steel construction material, a profile, formed with a specific shape or cross section and certain standards of chemical composition and mechanical properties. Structural steel shape, size, composition, strength, storage, etc., is regulated in most industrialized countries. Structural steel members, such as I-beams, have high second moments of area, which allow them to be very stiff in respect to their cross-sectional area.

### Common Structural Shapes

In most developed countries, the shapes available are set out in published standards, although a number of specialist and proprietary cross sections are also available.

- I-beam (I-shaped cross-section - in Britain these include Universal Beams (UB) and Universal Columns (UC); in Europe it includes the IPE, HE, HL, HD and other sections; in the US it includes Wide Flange (WF) and H sections)
- Z-Shape (half a flange in opposite directions)
- HSS-Shape (Hollow structural section also known as SHS (structural hollow section) and including square, rectangular, circular (pipe) and elliptical cross sections)
- Angle (L-shaped cross-section)
- Channel, or C section, ( [-shaped cross-section)
- Tee (T-shaped cross-section)
- Rail profile (asymmetrical I-beam)
  - Railway rail
  - Vignoles rail
  - Flanged T rail
  - Grooved rail
- Bar, a piece of metal, rectangular cross sectioned (flat) and long, but not so wide so as to be called a sheet.
- Rod, a round or square and long piece of metal or wood.
- Plate, metal sheets thicker than 6 mm or $^1/_4$ in.
- Open web steel joist

While many sections are made by hot or coldrolling, others are made by welding together flat or bent plates (for example, the largest circular hollow sections are made from flat plate bent into a circle and seam-welded).

## Standards

***Standard Structural Steels (Europe):*** Most steels used throughout Europe are specified to comply with the European standard EN 10025. However, many national standards also remain in force.

Typical grades are described as 'S275J2' or 'S355K2W'. In these examples, 'S' denotes structural rather than engineering steel; 275 or 355 denotes the yield strength in newtons per square millimetre or the equivalent megapascals; J2 or K2 denotes the materials toughness by reference to Charpy impact test values; and the 'W' denotes weathering steel. Further letters can be used to designate fine grain steel ('N' or 'NL'); quenched and tempered steel ('Q' or 'QL'); and thermomechanically rolled steel ('M' or 'ML').

The normal yield strength grades available are 195, 235, 275, 355, 420, and 460, although some grades are more commonly used than others e.g. in the UK, almost all structural steel is grades S275 and S355. Higher grades are available in quenched and tempered material (500, 550, 620, 690, 890 and 960 - although grades above 690 receive little if any use in construction at present).

A set of euronorms define the shape of a set of standard structural profiles:

- European I-beam: IPE - Euronorm 19-57
- European I-beam: IPN - DIN 1025-1
- European flange beams: HE - Euronorm 53-62
- European channels: UPN - DIN 1026-1
- European cold formed IS IS 800-1

### Standard Structural Steels (USA)

Steels used for building construction in the US use standard alloys identified and specified by ASTM International. These steels have an alloy identification beginning with A and then two, three, or four numbers. The four-number AISI steel grades commonly used for mechanical engineering, machines, and vehicles are a completely different specification series. The standard commonly used structural steels are:

*Carbon Steels:*

- A36 - structural shapes and plate
- A53 - structural pipe and tubing
- A500 - structural pipe and tubing
- A501 - structural pipe and tubing
- A529 - structural shapes and plate

*High Strength Low Alloy Steels:*

- A441 - structural shapes and plates
- A572 - structural shapes and plates
- A618 - structural pipe and tubing
- A992 - Possible applications are W or S I-Beams.
- A270 - structural shapes and plates

*Corrosion Resistant High Strength Low Alloy Steels:*

- A242 - structural shapes and plates
- A588 - structural shapes and plates

***Quenched and Tempered Alloy Steels:***

- A514 - structural shapes and plates
- A517 - boilers and pressure vessels

### CE Marking

The concept of CE marking for all construction products and steel products is introduced by the Construction Products Directive (CPD). The CPD is a European Directive that ensures the free movement of all construction products within the European Union.

Because steel components are "safety critical", CE Marking is not allowed unless the Factory Production Control (FPC) system under which they are produced has been assessed by a suitable certification body that has been approved to the European Commission.

In the case of steel products such as sections, bolts and fabricated steelwork the CE Marking demonstrates that the product complies with the relevant harmonized standard.

For steel structures the main harmonized standards are:

- Steel sections and plate - EN 10025-1
- Hollow sections - EN 10219-1 and EN 10210-1
- Pre-loadable bolts - EN 14399-1
- Non-preloadable bolts - EN 15048-1
- Fabricated steel - EN EN 1090 -1

The standard that covers CE Marking of structural steelwork is EN 1090-1. The standard has come into force in late 2010. After a transition period of two years, CE Marking will become mandatory in most European Countries sometime early in 2012. The official end date of the transition period is July 1, 2014.!

### Steel vs. Concrete

***Choosing the Ideal Structural Material:*** Most construction projects require the use of hundreds of different materials. These range from concrete of all different specifications, structural steel of different specifications, clay, mortar, ceramics, wood, etc. In terms of a load bearing structural frame, they will generally consist of structural steel, concrete, masonry, and/or wood, using an infinite combination of each to produce an efficient structure. Most commercial and industrial structures are primarily constructed using either structural steel or reinforced concrete. When designing a structure, an engineer must decide which, if not both, material is most suitable for the design. There are many factors considered when choosing a construction material. Cost is commonly the controlling element; however, other considerations such as weight, strength, constructability, availability, sustainability, and fire resistance will be taken into account before a final decision is made.

- Cost - The cost of these construction materials will depend entirely on the geographical location of the project and the availability of the materials. Just as price gasoline fluctuates, so does cement, aggregate, steel, etc. Reinforced concrete attributes about half of its construction costs to the required form-work. This refers to the lumber necessary to build the "box" or container to pour the concrete into place and remain there until it cures. Due to this high cost, precast concrete is becoming a popular option for designers so costs can be saved, as well as time. Structural steel costs will depend most entirely on the amount of steel used since steel is sold by the pound. It is therefore the job of the structural designer to specify the lightest steel members possible while still maintaining a safe structural design. Another method of reducing costs in steel design is to use many of the same size steel members, as opposed to many unique members.
- Strength/weight ratio - Construction materials are commonly categorized by their strength to weight ratio, or specific strength. This is defined as the strength of a material over its density. This gives an engineer an indication as to how useful the material is in comparison to its weight, with the weight being a direct indication of its cost (typically) and ease of construction. Concrete is typically ten times stronger in compression than in tension, giving it a higher strength to weight ratio in compression, only.
- Sustainability - Many construction companies and material vendors are making changes to be a more environmentally friendly company. Sustainability has become an entirely new consideration for materials that are to be placed into the environment for generations of time. A sustainable material will be one that has minimal effect on the environment, both at the time of installation as well as throughout the life cycle of the material. Reinforced concrete and structural steel both have the ability to be a sustainable construction option, if used properly. Over 80% of structural steel members fabricated today come from recycled metals, called A992 steel. This

member material is cheaper, as well as having a higher strength to weight ratio, than previously used steel members (A36 grade). Reinforced concrete can also be a very sustainable option. Concrete's material components are naturally occurring materials that are not harmful to the environment. Also, concrete can now be poured to be permeable, allowing the flow of water through a paved surface and therefore reducing the necessary infrastructural construction that causes further potential disturbance to the environment. Concrete can also be crushed and used as aggregate in future concrete applications, meaning it doesn't necessarily have to be used as a land fill.

- Fire resistance - One of the most dangerous hazards to a building is a fire hazard. This is especially true in dry, windy climates and for structures constructed using wood. Special considerations must be taken into account with structural steel to ensure it is not under a dangerous fire hazard condition. Reinforced concrete characteristically does not pose a threat in the event of fire and even resists the spreading of fire, as well as temperature changes. This makes concrete an excellent insulation, improving the sustainability of the building it surrounds by reducing the required energy to maintain climate.
- Corrosion - When choosing a structural material, it is important to consider the life cycle of the building. Some materials are susceptible to corrosion from their surrounding elements, such as water, heat, humidity, or salt. Special considerations must be taken into account during the installation of a structural material to prevent any potential corrosion hazards. This must also be made clear to the occupants of the building because there may or may not be a necessary maintenance requirement to prevent corrosion. For example, structural steel cannot be exposed to the environment because any moisture, or other contact with water, will cause it to rust. When the steel rusts it compromises the structural integrity of the building and poses a potential danger to the residual or surrounding occupants.

### Reinforced Concrete

- Characteristics - Generally consisting of portland cement, water, construction aggregate (coarse and fine), and steel reinforcing bars (rebar), concrete is cheaper in comparison to structural steel.

- Strength - Concrete is a composite material with relatively high compressive strength properties, but lacking in tensile strength/ductility. This inherently makes concrete a useful material for carrying the weight of a structure. Concrete reinforced with steel rebar give the structure a stronger tensile capacity, as well as an increase in ductility and elasticity.
- Constructability - Reinforced concrete must be poured and left to set, or harden. After setting (typically 1–2 days), a concrete must cure, the process in which concrete experiences a chemical reaction between the cementitious particles and the water. The curing process is complete after 28 days; however, construction may continue after 1–2 weeks, depending on the nature of the structure. Concrete can be constructed into nearly any shape and size. Approximately half of the cost of using reinforced concrete in a structural project is attributed to the construction of the form-work. In order to save time, and therefore costs, structural concrete members may be pre-cast. This refers to a reinforced concrete beam, girder, or column being poured off site and left to cure. After the curing process, the concrete member may be delivered to the construction site and installed as soon as it is needed. Since the concrete member was cured off location beforehand, construction may continue immediately after erection.
- Fire resistance - Concrete has excellent fire resistance properties, requiring no additional construction costs to adhere to the International Building Code (IBC) fire protection standards. However, concrete buildings will still likely use other materials that are not fire resistant. Therefore, a designer must still take into account the use of the concrete and where it will require fire hazardous materials in order to prevent future complications in the overall design.
- Corrosion - Reinforced concrete, when constructed properly, has excellent corrosion resistance properties. Concrete is not only resistant to water, but needs it to cure and develop its strength over time. However, the steel reinforcement in the concrete must not be exposed in order to prevent its corrosion as this could significantly reduce the ultimate strength of the structure. The American Concrete Institute provides the necessary design specifications for an engineer to ensure there is enough concrete covering any steel reinforcement to prevent exposure to water. This cover distance must be specified because

concrete will inevitable crack at locations carrying tension, or locations containing reinforcing bars for the purpose of carrying said tension. If the concrete cracks, it provides a path for water to travel directly to the reinforcing bars. Some reinforcing bars are coated in epoxy as a second order measure of preventing corrosion due to water contact. This method induces higher costs on the overall project, however, due to the higher cost of the epoxy coated bars. Also, when using epoxy coated bars, reinforced concrete members must be designed larger, as well as stronger, in order to balance the loss of friction between the reinforcing bars and concrete. This friction is referred to as bond strength, and it is vital to the structural integrity of a concrete member.

### *Structural Steel*

- Characteristics - Structural steel differs from concrete in its attributed compressive strength as well as tensile strength.
- Strength - Having high strength, stiffness, toughness, and ductile properties, structural steel is one of the most commonly used materials in commercial and industrial building construction.
- Constructability - Structural steel can be developed into nearly any shape, which are either bolted or welded together in construction. Structural steel can be erected as soon as the materials are delivered on site, whereas concrete must be cured at least 1–2 weeks after pouring before construction can continue, making steel a schedule-friendly construction material.
- Fire resistance - Steel is inherently a noncombustible material. However, when heated to temperatures seen in a fire scenario, the strength and stiffness of the material is significantly reduced. The International Building Code requires steel be enveloped in sufficient fire-resistant materials, increasing overall cost of steel structure buildings.
- Corrosion - Steel, when in contact with water, can corrode, creating a potentially dangerous structure. Measures must be taken in structural steel construction to prevent any lifetime corrosion. The steel can be painted, providing water resistance. Also, the fire resistance material used to envelope steel is commonly water resistant.

The tallest structures today (commonly called "skyscrapers" or high-rise) are constructed using structural steel due to its constructability, as well as its high strength-to-weight ratio. In comparison, concrete, while being less dense than steel, has a much lower strength-to-weight ratio. This is due to the much larger volume required for a structural concrete member to support the same load; steel, though denser, does not require as much material to carry a load. However, this advantage becomes insignificant for low-rise buildings, or those with several stories or less. Low-rise buildings distribute much smaller loads than high-rise structures, making concrete the economical choice. This is especially true for simple structures, such as parking garages, or any building that is a simple, rectilinear shape.

Structural steel and reinforced concrete are not always chosen solely because they are the most ideal material for the structure. Companies rely on the ability to turn a profit for any construction project, as do the designers. The price of raw materials (steel, cement, coarse aggregate, fine aggregate, lumber (for form-work), etc., is constantly changing. If a structure could be constructed using either material, the cheapest of the two will likely control. Another significant variable is the location of the project. The closest steel fabrication facility may be much further from the construction site than the nearest concrete supplier. The high cost of energy and transportation will control the selection of the material as well. All of these costs will be taken into consideration before the conceptual design of a construction project is begun.

### Combining Steel and Reinforced Concrete

Structures consisting of both materials utilise the benefits of structural steel and reinforced concrete. This is already common practice in reinforced concrete in that the steel reinforcement is used to provide steel's tensile strength capacity to a structural concrete member. A commonly seen example would be parking garages. Some parking garages are constructed using structural steel columns and reinforced concrete slabs. The concrete will be poured for the foundational footings, giving the parking garage a surface to be built on. The steel columns will be connected to the slab by bolting and/or welding them to steel studs extruding from the surface of the poured concrete slab. Pre-cast concrete beams may be delivered on site to be installed for the second floor, after which a concrete slab may be poured for the pavement area. This can be done for multiple stories. A parking garage of this type is just one possible example of many structures that may use both reinforced concrete and structural steel.

A structural engineer understands that there are an infinite number of designs that will produce an efficient, safe, and affordable building. It is the engineer's job to work alongside the owner(s), contractor(s), and all other parties involved to produce an ideal product that suits everyone's needs. When choosing the structural materials for their structure, the engineer has many variables to consider, such as the cost, strength/weight ratio, sustainability of the material, constructability, etc.

### Thermal Properties

The properties of steel vary widely, depending on its alloying elements. The austenizing temperature, the temperature where a steel transforms to an austenite crystal structure, for steel starts at 900 °C (1,650 °F) for pure iron, then, as more carbon is added, the temperature falls to a minimum 724 °C (1,335 °F) for eutectic steel (steel with only .83% by weight of carbon in it). As 2.1% carbon (by mass) is approached, the austenizing temperature climbs back up, to 1,130 °C (2,070 °F). Similarly, the melting point of steel changes based on the alloy.

The lowest temperature at which a plain carbon steel can begin to melt, its solidus, is 1,130 °C (2,070 °F). Steel never turns into a liquid below this temperature. Pure Iron ('Steel' with 0% Carbon) starts to melt at 1,492 °C (2,718 °F), and is completely liquid upon reaching 1,539 °C (2,802 °F). Steel with 2.1% Carbon by weight begins melting at 1,130 °C (2,070 °F), and is completely molten upon reaching 1,315 °C (2,399 °F). 'Steel' with more than 2.1% Carbon is no longer Steel, but is known as Cast iron.

### Fire Resistance

Steel loses strength when heated sufficiently. The *critical temperature* of a steel member is the temperature at which it cannot safely support its load. Building codes and structural engineering standard practice defines different critical temperatures depending on the structural element type, configuration, orientation, and loading characteristics. The critical temperature is often considered the temperature at which its yield stress has been reduced to 60% of the room temperature yield stress. In order to determine the fire resistance rating of a steel member, accepted calculations practice can be used, or a fire test can be performed, the critical temperature of which is set by the standard accepted to the Authority Having Jurisdiction, such as a building code. In Japan, this is below 400°C. In China, Europe and North America (e.g., ASTM E-119), this is approximately

1000–1300F (530-810C). The time it takes for the steel element that is being tested to reach the temperature set by the test standard determines the duration of the fire-resistance rating. Heat transfer to the steel can be slowed by the use of fireproofing materials, thus limiting steel temperature. Common fireproofing methods for structural steel include intumescent, endothermic and plaster coatings as well as drywall, calcium silicate cladding, and mineral or high temperature insulation mineral wool blanket.

Concrete building structures often meet code required fire-resistance ratings, as the concrete thickness over the steel rebar provides sufficient fire resistance. However, concrete can be subject to spalling, particularly if it has an elevated moisture content. Although additional fireproofing is not often applied to concrete building structures, it is sometimes used in traffic tunnels and locations where a hydrocarbon fuel fire is more likely, as flammable liquid fires provides more heat to the structural element as compared to a fire involving ordinary combustibles during the same fire period. Structural steel fireproofing materials include intumescent, endothermic and plaster coatings as well as drywall, calcium silicate cladding, and mineral or high temperature insulation wool blankets. Attention is given to connections, as the thermal expansion of structural elements can compromise fire-resistance rated assemblies.

### Manufacturing

Cutting workpieces to length is usually done with a bandsaw. A beam drill line (drill line) has long been considered an indispensable way to drill holes and mill slots into beams, channels and HSS elements. CNC beam drill lines are typically equipped with feed conveyors and position sensors to move the element into position for drilling, plus probing capability to determine the precise location where the hole or slot is to be cut.

For cutting irregular openings or non-uniform ends on dimensional (non-plate) elements, a cutting torch is typically used. Oxy-fuel torches are the most common technology and range from simple hand-held torches to automated CNC coping machines that move the torch head around the structural element in accordance with cutting instructions programmed into the machine.

Fabricating flat plate is performed on a plate processing centre where the plate is laid flat on a stationary 'table' and different cutting heads traverse the plate from a gantry-style arm or "bridge." The cutting heads can include a punch, drill or torch.

## I-Beam

An I-beam, also known as H-beam, W-beam (for "wide flange"), Universal Beam (UB), Rolled Steel Joist (RSJ), or double-T (especially in Polish, Bulgarian, Spanish, Italian and German), is a beam with an I- or H-shaped cross-section. The horizontal elements of the "I" are flanges, while the vertical element is termed the "web".

The web resists shear forces, while the flanges resist most of the bending moment experienced by the beam. Beam theory shows that the I-shaped section is a very efficient form for carrying both bending and shear loads in the plane of the web. On the other hand, the cross-section has a reduced capacity in the transverse direction, and is also inefficient in carrying torsion, for which hollow structural sections are often preferred.

### History

The method of producing an I-beam, as rolled from a single piece of steel, was patented by Alphonse Halbou of the company *Forges de la Providence* in 1849.

Bethlehem Steel was a leading supplier of rolled structural steel of various cross-sections in American bridge and skyscraper work of the mid-twentieth century. Today, rolled cross-sections have been partially displaced in such work by fabricated cross-sections.

### Overview

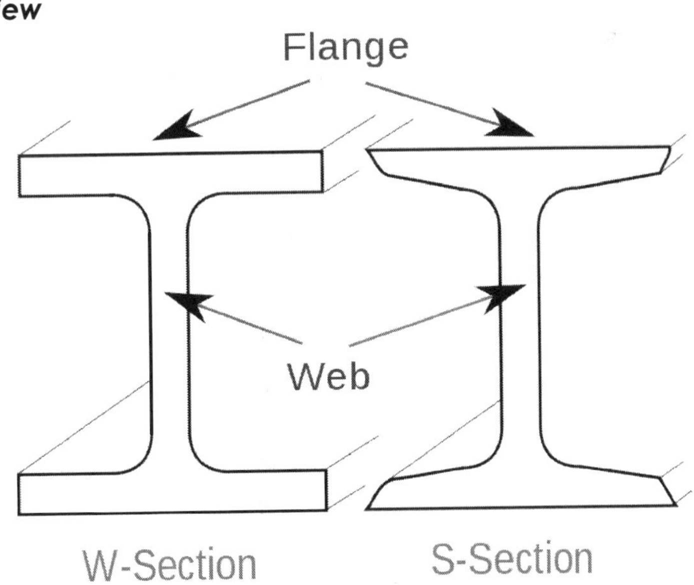


***Figure:*** *Typical cross-sections of I-beams.*

There are two standard I-beam forms:

- Rolled I-beam, formed by hot rolling, cold rolling or extrusion (depending on material).
- Plate girder, formed by welding (or occasionally bolting or riveting) plates.

I-beams are commonly made of structural steel but may also be formed from aluminium or other materials. A common type of I-beam is the *rolled steel joist* (RSJ)—sometimes incorrectly rendered as *reinforced steel joist*. British and European standards also specify Universal Beams (UBs) and Universal Columns (UCs). These sections have parallel flanges, as opposed to the varying thickness of RSJ flanges which are seldom now rolled in the UK. Parallel flanges are easier to connect to and do away with the need for tapering washers. UCs have equal or near-equal width and depth and are more suited to being orientated vertically to carry axial load such as columns in multi-storey construction, while UBs are significantly deeper than they are wide are more suited to carrying bending load such as beam elements in floors.

I-joists — I-beams engineered from wood with fibreboard and/or laminated veneer lumber — are also becoming increasingly popular in construction, especially residential, as they are both lighter and less prone to warping than solid wooden joists. However there has been some concern as to their rapid loss of strength in a fire if unprotected.

### Design

I-beams are widely used in the construction industry and are available in a variety of standard sizes. Tables are available to allow easy selection of a suitable steel I-beam size for a given applied load. I-beams may be used both as beams and as columns.

I-beams may be used both on their own, or acting compositely with another material, typically concrete. Design may be governed by any of the following criteria:

- deflection: the stiffness of the I-beam will be chosen to minimize deformation
- vibration: the stiffness and mass are chosen to prevent unacceptable vibrations, particularly in settings sensitive to vibrations, such as offices and libraries
- bending failure by yielding: where the stress in the cross section exceeds the yield stress

- bending failure by lateral torsional buckling: where a flange in compression tends to buckle sideways or the entire cross-section buckles torsionally
- bending failure by local buckling: where the flange or web is so slender as to buckle locally
- local yield: caused by concentrated loads, such as at the beam's point of support
- shear failure: where the web fails. Slender webs will fail by buckling, rippling in a phenomenon termed tension field action, but shear failure is also resisted by the stiffness of the flanges
- buckling or yielding of components: for example, of stiffeners used to provide stability to the I-beam's web.

### Design for Bending

A beam under bending sees high stresses along the axial fibres that are farthest from the neutral axis. To prevent failure, most of the material in the beam must be located in these regions. Comparatively little material is needed in the area close to the neutral axis. This observation is the basis of the I-beam cross-section; the neutral axis runs along the centre of the web which can be relatively thin and most of the material can be concentrated in the flanges.

The ideal beam is the one with the least cross-sectional area (and hence requiring the least material) needed to achieve a given section modulus. Since the section modulus depends on the value of the moment of inertia, an efficient beam must have most of its material located as far from the neutral axis as possible. The farther a given amount of material is from the neutral axis, the larger is the section modulus and hence a larger bending moment can be resisted.

When designing a symmetric I-beam to resist stresses due to bending the usual starting point is the required section modulus. If the allowable stress is $\sigma_{max}$ and the maximum expected bending moment is $M_{max}$, then the required section modulus is given by

$$S = \frac{M_{max}}{\sigma_{max}} = \frac{I}{c}$$

where $I$ is the moment of inertia of the beam cross-section and $c$ is the distance of the top of the beam from the neutral axis. For a beam of cross-sectional area $a$ and height $h$, the ideal cross-section would have half the area at a distance $h/2$ above the cross-section and the other half at a distance $h/2$ below the cross-section For this cross-section

$$I = \frac{ah^2}{4};\ S = 0.5ah\ .$$

However, these ideal conditions can never be achieved because material is needed in the web for physical reasons, including to resist buckling. For wide-flange beams, the section modulus is approximately

$$S \approx 0.35ah$$

which is superior to that achieved by rectangular beams and circular beams.

### Issues

Though I-beams are excellent for unidirectional bending in a plane parallel to the web, they do not perform as well in bidirectional bending. These beams also show little resistance to twisting and undergo sectional warping under torsional loading. For torsion dominated problems, box beams and other types of stiff sections are used in preference to the I-beam.

### Wide-Flange Steel Materials and Rolling Processes (U.S.)

In the United States, the most commonly mentioned I-beam is the wide-flange (W) shape. These beams have flanges in which the planes are nearly parallel. Other I-beams include American Standard (designated S) shapes, in which flange surfaces are not parallel, and H-piles (designated HP), which are typically used as pile foundations. Wide-flange shapes are available in grade ASTM A992, which has generally replaced the older ASTM grades A572 and A36. Ranges of yield strength:

- A36: 36,000 psi (250 MPa)
- A572: 42,000–60,000 psi (290–410 MPa), but 50,000 psi (340 MPa) is the most common
- A588: Similar to A572
- A992: 50,000–65,000 psi (340–450 MPa)

Like most steel products, I-beams often contain some recycled content. The American Institute of Steel Construction ("AISC") publishes the "Steel Construction Manual" for designing structures of various shapes. It documents the common approaches, ASD and LRFD, (as of 13th ed.) to creating such designs.

### Standards

The following standards define the shape and tolerances of I-beam steel sections:

### Euronorms

- EN 10024, Hot rolled taper flange I sections - Tolerances on shape and dimensions.
- EN 10034, Structural steel I and H sections - Tolerances on shape and dimensions.
- EN 10162, Cold rolled steel sections - Technical delivery conditions - Dimensional and cross-sectional tolerances

### Other

- DIN 1025-5
- ASTM A6, American Standard Beams
- BS 4-1
- IS 808 - Dimensions hot rolled steel beam, column, channel and angle sections

### Designation and Terminology

- In the United States, steel I-beams are commonly specified using the depth and weight of the beam. For example, a "W10x22" beam is approximately 10 inches (25 cm) in depth (nominal height of the I-beam from the outer face of one flange to the outer face of the other flange) and weighs approximately 22 lb/ft (33 kg/m). It should be noted that wide flange section often vary from their nominal depth. In the case of the W14 series, they may be as deep as 22.84 inches (58.0 cm).
- In Canada, steel I-beams are now commonly specified using the depth and weight of the beam in metric terms. For example, a "W250x33" beam is approximately 250 millimetres (9.8 in) in depth (height of the I-beam from the outer face of one flange to the outer face of the other flange) and weighs approximately 33 kg/m (22 lb/ft). I-beams are still available in U.S. sizes from many Canadian manufacturers.
- In IndiaI-beams are designated as ISMB, ISJB, ISLB, ISWB. ISMB: Indian Standard Medium Weight Beam, ISJB: Indian Standard Junior Beams, ISLB: Indian Standard Light Weight Beams, and ISWB: Indian Standard Wide Flange Beams. Beams are designated as per respective abbreviated reference followed by the depth of section, such as for example "ISMB 450", where 450 is the depth of section in millimetres (mm). The dimensions of these beams are classified as per IS:808 (as per BIS).
- In the United Kingdom, these steel sections are commonly specified with a code consisting of the major dimension (usually

the depth)-x-the minor dimension-x-the mass per metre-ending with the section type, all measurements being metric. Therefore a 152x152x23UC would be a column section (UC = universal column) of approximately 152 millimetres (6.0 in) depth 152 mm width and weighing 23 kilograms per metre (46 lb/yd) of length.

- In Australia, these steel sections are commonly referred to as Universal Beams (UB) or Columns (UC). The designation for each is given as the height of the beam, the type (beam or column) and then the unit metre rate (e.g., a 460UB67 is a 460 millimetres (18.1 in) deep universal beam that weighs 67 kilograms per metre (135 lb/yd)).

### Indian Standard Beams ISMB

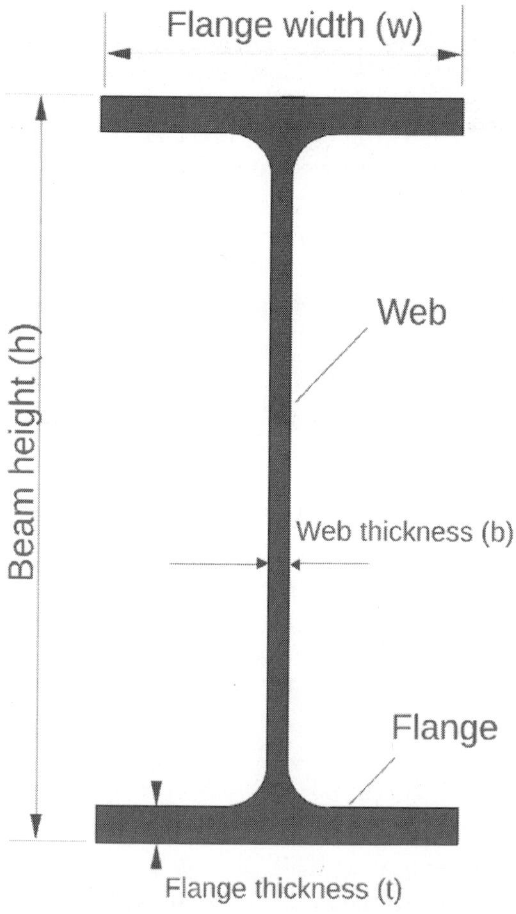


***Figure:*** *Wide-flange I-beam.*

| *Type* | *Beam height (mm)* | *Flange width (mm)* | *Web thickness (mm)* | *Flange thickness (mm)* | *Weight (kg/m)* | *Cross-section area ($cm^2$)* | *Moment of inertia in torsion (J) ($cm^4$)* |
|---|---|---|---|---|---|---|---|
| ISMB 80 | 80 | 46 | 3.8 | 5.2 | 6.0 | 7.64 | 0.70 |
| ISMB 100 | 100 | 55 | 4.1 | 5.7 | 8.1 | 10.3 | 1.10 |
| ISMB 120 | 120 | 70 | 4.4 | 6.3 | 10.4 | 13.2 | 1.71 |
| ISMB 140 | 140 | 73 | 4.7 | 6.9 | 12.9 | 16.4 | 2.54 |
| ISMB 750 × 137 | 753 | 263 | 11.5 | 17 | 137 | 175 | 137.1 |
| ISMB 750 × 147 | 753 | 265 | 13.2 | 17 | 147 | 188 | 161.5 |
| ISMB 750 × 173 | 762 | 267 | 14.4 | 21.6 | 173 | 221 | 273.6 |
| ISMB 750 × 196 | 770 | 268 | 15.6 | 25.4 | 196 | 251 | 408.9 |

## European Wide Flange Beams HEA and HEB

| *Type* | *Beam height (mm)* | *Flange width (mm)* | *Web thickness (mm)* | *Flange thickness (mm)* | *Weight (kg/m)* | *Cross-section area ($cm^2$)* | *Moment of inertia in torsion (J) ($cm^4$)* |
|---|---|---|---|---|---|---|---|
| HE 100 A | 96 | 100 | 5 | 8 | 16.7 | 21.2 | 5.24 |
| HE 120 A | 114 | 120 | 5 | 8 | 19.9 | 25.3 | 5.99 |
| HE 140 A | 133 | 140 | 5.5 | 8.5 | 24,7 | 31.4 | 8.13 |
| HE 160 A | 152 | 160 | 6 | 9 | 30.4 | 38.8 | 12.19 |
| HE 1000 × 415 | 1020 | 304 | 26 | 46 | 415 | 528.7 | 2714 |
| HE 1000 × 438 | 1026 | 305 | 26.9 | 49 | 437 | 557.2 | 3200 |
| HE 1000 × 494 | 1036 | 309 | 31 | 54 | 494 | 629.1 | 4433 |
| HE 1000 × 584 | 1056 | 314 | 35.6 | 64 | 584 | 743.7 | 7223 |

## Cellular Beams

Cellular beams are the modern version of the traditional "castellated" beam which results in a beam approximately 40–60% deeper than its parent section. The exact finished depth, cell diameter and cell spacing are flexible. A cellular beam is up to 1.5 times stronger than its parent section and is therefore utilised to create efficient large span constructions.

## Column

Column or pillar in architecture and structural engineering is a structural element that transmits, through compression, the weight of the structure above to other structural elements below. In other words, a column is a compression member. The term column applies especially to a large round support with a capital and base and made of stone, or appearing to be so. A small wooden or metal support is typically called a post, and supports with a rectangular or other non-round section are usually called piers. For the purpose of wind or earthquake engineering, columns may be designed to resist lateral forces. Other compression members are often termed "columns" because of the similar stress conditions. Columns are frequently used to support beams or arches on which the upper parts of walls or ceilings rest. In architecture, "column" refers to such a structural element that also has certain proportional and decorative features. A column might also be a decorative element not needed for structural purposes; many columns are "engaged", that is to say form part of a wall.

### History

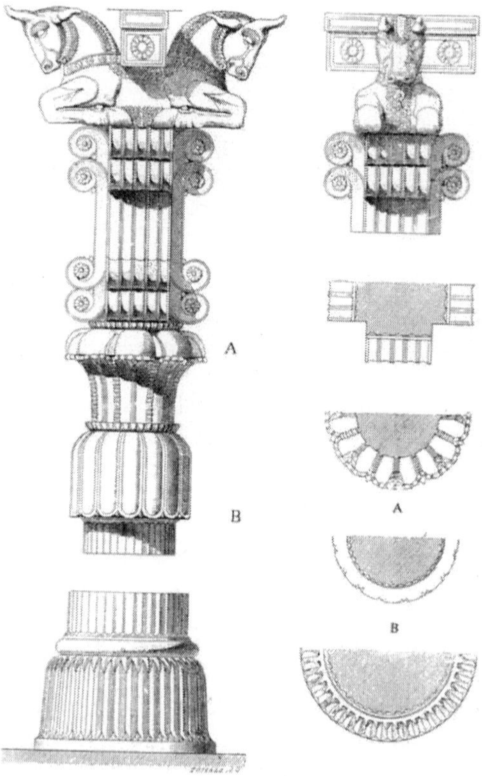


***Figure:*** *Plan, Front view and side view of a typical Persepolis column, of Persia (Iran).*

All significant Iron Age civilizations of the Near East and Mediterranean made some use of columns. In Ancient Egyptian architecture as early as 2600 BC the architect Imhotep made use of stone columns whose surface was carved to reflect the organic form of bundled reeds; in later Egyptian architecture faceted cylinders were also common.

Some of the most elaborate columns in the ancient world were those of the Persians, especially the massive stone columns erected in Persepolis. They included double-bull structures in their capitals. The Hall of Hundred Columns at Persepolis, measuring 70 × 70 metres, was built by the Achaemenid king Darius I (524–486 BC). Many of the ancient Persian columns are standing, some being more than 30 metres tall.

The Egyptians, Persians and other civilizations mostly used columns for the practical purpose of holding up the roof inside a building, preferring outside walls to be decorated with reliefs or painting, but the Ancient Greeks, followed by the Romans, loved to use them on the outside as well, and the extensive use of columns on the interior and exterior of buildings is one of the most characteristic features of classical architecture, in buildings like the Parthenon. The Greeks developed the classical orders of architecture, which are most easily distinguished by the form of the column and its various elements. Their Doric, Ionic, and Corinthian orders were expanded by the Romans to include the Tuscan and Composite orders.

Columns, or at least large structural exterior ones, became much less significant in the architecture of the Middle Ages, and the classical forms were abandoned in both Byzantine architecture and the Romanesque and Gothic architecture or Europe in favour of more flexible forms, with capitals often using various types of foliage decoration, and in the West scenes with figures carved in relief. Renaissance architecture was keen to revive the classical vocabulary and styles, and the informed use and variation of the classical orders remained fundamental to the training of architects throughout Baroque, Rococo and Neo-classical architecture.

### Structure

Early columns were constructed of stone, some out of a single piece of stone. Monolithic columns are among the heaviest stones used in architecture. Other stone columns are created out of multiple sections of stone, mortared or dry-fit together. In many classical sites, sectioned columns were carved with a centre hole or depression so that they could be pegged together, using stone or metal pins. The design of most classical columns incorporates entasis (the inclusion of a slight outward curve in the sides) plus a reduction in diameter along the height of the column, so that the top is as little as 83% of the bottom diameter. This reduction mimics the parallax effects which the eye expects to see, and tends to make columns look taller and straighter than they are while entasis adds to that effect.

### Nomenclature

Classical columns have a bottom or plinth that rests on the stylobate or foundation. At the top a capital (also known as a chapiter) or finial decorate the column, depending on whether it is attached to a structure or free-standing.

Modern columns are constructed out of steel, poured or precast concrete, or brick. They may then be clad in an architectural covering (or veneer), or left bare. In modern terms the impost (or pier) is the topmost member of a column. The bottom-most part of the arch, called the springing, rests on the impost.

### Equilibrium, Instability, and Loads

***Mechanical Failure Modes:*** As the axial load on a perfectly straight slender column with elastic material properties is increased in magnitude, this ideal column passes through three states: stable equilibrium, neutral equilibrium, and instability. The straight column under load is in stable equilibrium if a lateral force, applied between the two ends of the column, produces a small lateral deflection which disappears and the column returns to its straight form when the lateral force is removed. If the column load is gradually increased, a condition is reached in which the straight form of equilibrium becomes so-called neutral equilibrium, and a small lateral force will produce a deflection that does not disappear and the column remains in this slightly bent form when the lateral force is removed. The load at which neutral equilibrium of a column is reached is called the critical or buckling load. The state of instability is reached when a slight increase of the column load causes uncontrollably growing lateral deflections leading to complete collapse.

For an axially loaded straight column with any end support conditions, the equation of static equilibrium, in the form of a differential equation, can be solved for the deflected shape and critical load of the column. With hinged, fixed or free end support conditions the deflected shape in neutral equilibrium of an initially straight column with uniform cross section throughout its length always follows a partial or composite sinusoidal curve shape, and the critical load is given by

$$f_{cr} \equiv \frac{\pi^2 EI_{min}}{L^2} \tag{1}$$

where $E$ = elastic modulus of the material, $I_{min}$ = the minimal moment of inertia of the cross section, and $L$ = actual length of the column between its two end supports. A variant of (1) is given by

$$f_{cr} \equiv \frac{\pi^2 E_T}{(\frac{KL}{r})^2} \tag{2}$$

where $r$ = radius of gyration of [column]cross-section which is equal to the square root of (I/A), $K$ = ratio of the longest half sine wave to the actual column length, and $KL$ = effective length (length of an equivalent hinged-hinged column). From Equation (2) it can be noted that the buckling strength of a column is inversely proportional to the square of its length.

When the critical stress, $F_{cr}$ ($F_{cr} = P_{cr}/A$, where $A$ = cross-sectional area of the column), is greater than the proportional limit of the material, the column is experiencing inelastic buckling. Since at this stress the slope of the material's stress-strain curve, $E_t$ (called the tangent modulus), is smaller than that below the proportional limit, the critical load at inelastic buckling is reduced.

More complex formulas and procedures apply for such cases, but in its simplest form the critical buckling load formula is given as Equation (3),

$$f_{cr} \equiv F_y - \frac{F_y^2}{4\pi^2 E}\left(\frac{KL}{r^2}\right) \tag{3}$$

where $E_t$ = tangent modulus at the stress $F_{cr}$

A column with a cross section that lacks symmetry may suffer torsional buckling (sudden twisting) before, or in combination with, lateral buckling. The presence of the twisting deformations renders both theoretical analyses and practical designs rather complex.

Eccentricity of the load, or imperfections such as initial crookedness, decreases column strength. If the axial load on the column is not concentric, that is, its line of action is not precisely coincident with the centroidal axis of the column, the column is characterized as eccentrically loaded. The eccentricity of the load, or an initial curvature, subjects the column to immediate bending. The increased stresses due to the combined axial-plus-flexural stresses result in a reduced load-carrying ability.

Column elements are considered to be massive if minimal side dimension is equal or more than 400 mm. Massive columns have ability to increase concrete strength during long time period (even during exploitation period). Taking into account possible loads onto structure increase in future (and even threat of progressive failure, terroristic attacks, explosions etc.) massive columns have advantage comparing with not ones. A little economy today has no sense as usual for future.

Moreover relatively small sections are not technological for reinforced structures during their production. Balance between economy, mass of structures and so called "sustainable" construction is necessary.

### Extensions

When a column is too long to be built or transported in one piece, it has to be extended or spliced at the construction site. A reinforced concrete column is extended by having the steel reinforcing bars protrude a few inches or feet above the top of the concrete, then placing the next level of reinforcing bars to overlap, and pouring the concrete of the next level. A steel column is extended by welding or bolting splice plates on the flanges and webs or walls of the columns to provide a few inches or feet of load transfer from the upper to the lower column section. A timber column is usually extended by the use of a steel tube or wrapped-around sheet-metal plate bolted onto the two connecting timber sections.

when $e<12$ then it is known as short column

### Foundations

A column that carries the load down to a foundation must have means to transfer the load without overstressing the foundation material. Reinforced concrete and masonry columns are generally built directly on top of concrete foundations. A steel column, when seated on a concrete foundation, must have a base plate to spread the load over a larger area and thereby reduce the bearing pressure. The base plate is a thick rectangular steel plate usually welded to the bottom end of the column.

### Classical Orders

The Roman author Vitruvius, relying on the writings (now lost) of Greek authors, tells us that the ancient Greeks believed that their Doric order developed from techniques for building in wood in which the earlier smoothed tree trunk was replaced by a stone cylinder.

### Doric Order

The Doric order is the oldest and simplest of the classical orders. It is composed of a vertical cylinder that is wider at the bottom. It generally has neither a base nor a detailed capital. It is instead often topped with an inverted frustum of a shallow cone or a cylindrical band of carvings. It is often referred to as the masculine order because it is represented in the bottom level of the Colosseum and the Parthenon, and was therefore considered to be able to hold more weight. The height-to-thickness ratio is about 8:1. The shaft of a Doric Column is always fluted.

The Greek Doric, developed in the western Dorian region of Greece, is the heaviest and most massive of the orders. It rises from the stylobate without any base; it is from four to six times as tall as its diameter; it has twenty broad flutes; the capital consists simply of a banded necking swelling out into a smooth echinus, which carries a flat square abacus; the Doric entablature is also the heaviest, being about one-fourth the height column. The Greek Doric order was not used after c. 100 B.C. until its "rediscovery" in the mid-eighteenth century.

### Tuscan Order

The Tuscan order, also known as Roman Doric, is also a simple design, the base and capital both being series of cylindrical disks of alternating diameter. The shaft is almost never fluted. The proportions vary, but are generally similar to Doric columns. Height to width ratio is about 7:1.

### Ionic Order

The Ionic column is considerably more complex than the Doric or Tuscan. It usually has a base and the shaft is often fluted (it has grooves carved up its length). On the top is a capital in the characteristic shape of a scroll, called a volute, or scroll, at the four corners. The height-to-thickness ratio is around 9:1. Due to the more refined proportions and scroll capitals, the Ionic column is sometimes associated with academic buildings. Ionic style columns were used on the second level of the Colosseum.

### Corinthian Order

The Corinthian order is named for the Greek city-state of Corinth, to which it was connected in the period. However, according to the architectural historian Vitruvius, the column was created by the sculptor Callimachus, probably an Athenian, who drew acanthus leaves growing around a votive basket. In fact, the oldest known Corinthian capital was found in Bassae, dated at 427 BC. It is sometimes called the feminine order because it is on the top level of the Colosseum and holding up the least weight, and also has the slenderest ratio of thickness to height. Height to width ratio is about 10:1.

### Composite Order

The Composite order draws its name from the capital being a composite of the Ionic and Corinthian capitals. The acanthus of the Corinthian column already has a scroll-like element, so the distinction is sometimes subtle. Generally the Composite is similar to the Corinthian in proportion and employment, often in the upper tiers of colonnades. Height to width ratio is about 11:1 or 12:1.

### Solomonic

A Solomonic column, sometimes called "barley sugar", begins on a base and ends in a capital, which may be of any order, but the shaft twists in a tight spiral, producing a dramatic, serpentine effect of movement. Solomonic columns were developed in the ancient world, but remained rare there. A famous marble set, probably 2nd century, was brought to St Peter's, Rome by Constantine I, and placed round the saint's shrine, and was thus familiar throughout the Middle Ages, by which time they were thought to have been removed from the Temple of Jerusalem. The style was used in bronze by Bernini for his spectacular St. Peter's baldachin, actually a ciborium (which displaced Constantine's columns), and thereafter became very popular with Baroque and Rococo church architects, above all in Latin America, where they were very often used, especially on a small scale, as they are easy to produce in wood by turning on a lathe (hence also the style's popularity for spindles on furniture and stairs).

## Joist

In architecture and engineering, a joist is one of the horizontal supporting members that run from wall to wall, wall to beam, or beam to beam to support a ceiling, roof, or floor. It may be made of oriented strand board, plywood, wood, steel, or concrete. Typically, a beam is bigger than a joist, which is often supported by beams and laid out in repetitive patterns.

The wider the span between the supporting structures, the deeper the joist will need to be if it is not to deflect under load. Lateral support also increases its strength. There are approved formulas for calculating the depth required and reducing the depth as needed; however, a rule of thumb for calculating the depth of a wooden floor joist for a residential property is half the span in feet plus two inches; for example, the joist depth required for a 14 foot span is 9 inches. Many steel joist manufacturers supply load tables in order to allow designers to select the proper joist sizes for their projects.

Engineered wood products such as I-joists gain strength from the depth of the floor or the height of each joist, as well as the larger bottom and top chords, as compared to standard dimensional lumber joists. A common saying regarding structural design is that "deeper is cheaper", referring to the more cost-effective design of a given structure by using deeper but more expensive joists, because fewer joists needed and longer spans are achieved, which more than makes up for the added cost of deeper joists.

Bandsill is another term for joist used by construction workers and home inspectors in the southeastern United States.

In traditional timber framing there may be a single set of joists which carry both a floor and ceiling called a single floor (single joist floor, single framed floor) or two sets of joists, one carrying the floor and another carrying the ceiling called a double floor (double framed floor). The term binding joist is sometimes used to describe beams at floor level running perpendicular to the ridge of a gable roof and joined to the intermediate posts. Joists which land on a binding joist are called bridging joists. A large beam in the ceiling of a room carrying joists is a summer beam. A ceiling joist may be installed flush with the bottom of the beam or sometimes below the beam. Joists left exposed and visible from below are called "naked flooring" or "articulated" (a modern U.S. term) and were typically planed smooth (wrought) and sometimes chamfered or beaded. Joists may join to their supporting beams in many ways: joists resting on top of the supporting beams are said to be "lodged"; dropped in using a butt cog joint (a type of lap joint), half-dovetail butt cog, or a half-dovetail lap joint. Joists may also be tenoned in during the raising with a soffit tenon or a tusk tenon (possibly with a housing). Joists can also be joined by being slipped into mortises after the beams are in place such as a chase mortise (pulley mortise), L-mortise, or "short joist". Also, in some Dutch-American work, ground level joists are be placed on a foundation and then a sill placed on top of the joists such as what Jack Sobon called an "inverted sill" or with a "plank sill". Joists can have different joints on either ends such as being tenoned on one end and lodged on the other end. A reduction in the under-side of cogged joist-ends may be square, sloped or curved. Typically joists do not tie the beams together, but sometimes they are pinned or designed to hold under tension. Joists on the ground floor were sometimes a pole (pole joist, half round joist, log joist. A round timber with one flat surface) and in barns long joists were sometimes supported on a sleeper (a timber not joined to but supporting other beams). Joists left out of an area form an opening called a "well" as in a stairwell or chimney-well. The joists forming the well are the heading joist (header) and trimming joist (trimmer). Trimmers take the name of the feature such as hearth trimmer, stair trimer, etc. Shortened joists are said to be crippled. The term rim joist is rare before the 1940s in America, it forms the edge of a floor. The outermost joist in half timber construction may be of a more durable species than the interior joists. In a barn, loose poles above the drive floor are called a scaffold. Between the joists, the area called a joist-bay, and above the ceiling in some old houses is material called pugging which was used to deaden sound, insulate, and resist the spread of fire.

## Deck (Building)

In architecture, a deck is a flat surface capable of supporting weight, similar to a floor, but typically constructed outdoors, often elevated from the ground, and usually connected to a building. The term is a generalization of decks as found on ships.

### *Functions and Materials*

Wood or timber "decking" can be used in a number of ways: as part of garden landscaping, to extend living areas of houses, and as an alternative to stone based features such as patios. Decks are made from treated lumber, composite lumber, composite material, Aluminium, Western red cedar, teak, mahogany, ipê and other hardwoods and recycled planks made from high-density polyethylene (HDPE), polystyrene (PS) and PET plastic as well as mixed plastics and wood fibre (often called "composite" lumber). Artificial decking products are often called "wood-plastic composites". These days, WPC's have more widely known by different brands like Trex, Azek, Ecornboard etc.

Historically, the softwoods used for decking were logged from old growth forests. These include Atlantic white cedar, redwood and Western red cedar (redcedar). Atlantic City built the first coastal boardwalk in the United States, originally constructed of Atlantic white cedar. However, it was not long before the commercial logging of this tree and clearing of cedar swamps in New Jersey caused a decline in the availability of decking. Atlantic City and New York City both switched to Western red cedar. By the 1960s, Western red cedar from the US was declining due to over-logging. More expensive Western red cedar was available from western Canada (British Columbia) but by then, pressure treated pine had become available.

But even with chemical treatments (such as chromated copper arsenate or CCA), pine decking is not as durable as cedars in an outdoor environment. Thus, many municipalities and homeowners are turning to hardwoods. Decks are often built from pressure treated wood. Pressure treated wood is long lasting and holds up to wet and icey weather conditions. Pressure treated wood however is treated with chemicals which have been known to be toxic. Slivers received from pressure treated wood most generally become infected. Pressure treated saw dust also contains toxins such as strychnine, also often used as rat poison. These toxins, when inhaled, can require hospitalization for both acute and chronic exposures.

Generally, hardwoods used for decking come from tropical forests. Much of the logging taking place to produce these woods, especially teak, mahogany and ipê, is occurring illegally, as outlined in numerous reports by environmental organisations such as Greenpeace, Friends of the Earth and Rainforest Relief. US tropical wood imports are rising, partly due to the demand for decking.

Due to environmental concerns, composite decking (a mixture of two materials, typically wood pulp and recycled material such as plastic bottles or plastic bags) have appeared on the market. Proponents of composite decking have touted this as a much needed development as this helps to curb logging of trees for new decks. However composite decking has been found to contain harmful chemicals, cannot be refurbished, and despite claims from decking companies, the composite deck still attracts molding.

Residential decks may contain spaces for BBQing, dining and seating. BBQ areas ideally should be situated near the patio door while out of the way from general foot traffic. Dining spaces will include patio tables, for a typical 6 person outdoor patio table building an area of 12' x 16' is ideal. If deck space is available, homeowners may choose to include a seating area for outdoor couches and benches.

### Construction

The deck of a house is generally a wooden platform built above the ground and connected to the main building. It is generally enclosed by a railing for safety. Access may be from the house through doors and from the ground via a stairway. Residential decks can be constructed over steep areas or rough ground that is otherwise unusable. Decks can also be covered by a canopy or pergola to control sunlight. Deck designs can be found in numerous books, do-it-yourself magazines and web sites, and from the USDA.

Larger buildings may also have decks on the upper floors of the building which may be open to the public as observation decks or a Skyrise greenery. A deck is also the surface used to construct a boardwalk over sand on barrier islands.

Laying deck or throwing deck refers to the act of placing and bolting down cold-formed steel beneath roofing and concrete floors. This is usually done by an ironworker, sometimes in conjunction with a cement mason or carpenter. It regarded as one of the most physically demanding jobs in the iron working industry.

In the UK the various ban on smoking in public buildings was expected lead to an increase in the use of timber decking for outdoor spaces where smokers can gather.

### Roof Deck

The roof deck is the roofing material layer between the primary structural components (trusses&joists) and either insulative layers or weatherproofing layers in a typical roof system. Usually the roof deck acts as a unifying structural diaphragm by tying all the structural components together. There are many types of roof deck: plywood or OSB sheathing, wood tongue and groove, corrugated metal, grancrete encapsulated polystyrene, reinforced concrete, and double tee.

## Framing (Construction)

Framing, in construction known as light-frame construction, is a building technique based around vertical structural members, usually called studs, which provide a stable frame to which interior and exterior wall coverings are attached, and covered by a roof made of horizontal ceiling joists and sloping rafters (or pre-fabricated roof trusses).

Modern light-frame structures usually gain strength from rigid panels (plywood and other plywood-like composites such as oriented strand board (OSB) used to form all or part of wall sections) but until recently carpenters employed various forms of diagonal bracing (called *wind braces*) to stabilize walls. Diagonal bracing remains a vital interior part of many roof systems, and in-wall wind braces are required by building codes in many municipalities or by individual state laws in the United States.

Light frame construction using standardized dimensional lumber has become the dominant construction method in North America and Australia because of its economy. Use of minimal structural materials allows builders to enclose a large area with minimal cost, while achieving a wide variety of architectural styles. The ubiquitous platform framing and the older balloon framing are the two different light frame construction systems used in North America.

### Walls

Wall framing in house construction includes the vertical and horizontal members of exterior walls and interior partitions, both of bearing walls and non-bearing walls. These *stick* members, referred to as studs, wall plates and lintels (*headers*), serve as a nailing base for all covering material and support the upper floor platforms, which provide the lateral strength along a wall. The platforms may be the boxed structure of a ceiling and roof, or the ceiling and floorjoists of the story above. The technique is variously referred to colloquially in the building trades as *stick and frame, stick and platform*, or *stick and box* as the sticks (studs) give the structure its vertical support, and the box-shaped floor sections with joists contained within length-long post and lintels (more commonly called *headers*), support the weight of whatever is above, including the next wall up and the roof above the top story. The platform also provides the lateral support against wind and holds the stick walls true and square. Any lower platform supports the weight of the platforms and walls above the level of its component headers and joists.

Framing lumber should be grade-stamped, and have a moisture content not exceeding 19%.

There are three historically common methods of framing a house.

- Post and beam, which is now used predominately in barn construction.
- Balloon framing using a technique suspending floors from the walls was common until the late 1940s, but since that time, platform framing has become the predominant form of house construction.
- Platform framing often forms wall sections horizontally on the sub-floor prior to erection, easing positioning of studs and increasing accuracy while cutting the necessary manpower. The top and bottom plates are end-nailed to each stud with two nails at least 3.25 in (83 mm) in length (*16d* or *16 penny*

nails). Studs are at least doubled (creating posts) at openings, the jack stud being cut to receive the lintels (headers) that are placed and end-nailed through the outer studs.

Wall sheathing, usually a plywood or other laminate, is usually applied to the framing prior to erection, thus eliminating the need to scaffold, and again increasing speed and cutting manpower needs and expenses. Some types of exterior sheathing, such as asphalt-impregnated fibreboard, plywood, oriented strand board and waferboard, will provide adequate bracing to resist lateral loads and keep the wall square. (Construction codes in most jurisdictions require a stiff plywood sheathing.) Others, such as rigid glass-fibre, asphalt-coated fibreboard, polystyrene or polyurethane board, will not. In this latter case, the wall should be reinforced with a diagonal wood or metal bracing inset into the studs. In jurisdictions subject to strong wind storms (hurricane countries, tornado alleys) local codes or state law will generally require both the diagonal wind braces and the stiff exterior sheathing regardless of the type and kind of outer weather resistant coverings.

### *Corners*

A multiple-stud post made up of at least three studs, or the equivalent, is generally used at exterior corners and intersections to secure a good tie between adjoining walls and to provide nailing support for the interior finish and exterior sheathing. Corners and intersections, however, must be framed with at least two studs.

Nailing support for the edges of the ceiling is required at the junction of the wall and ceiling where partitions run parallel to the ceiling joists. This material is commonly referred to as 'dead wood' or backing.

### *Exterior Wall Studs*

Wall framing in house construction includes the vertical and horizontal members of exterior walls and interior partitions. These members, referred to as studs, wall plates and lintels, serve as a nailing base for all covering material and support the upper floors, ceiling and roof.

Exterior wall studs are the vertical members to which the wall sheathing and cladding are attached. They are supported on a bottom plate or foundation sill and in turn support the top plate. Studs usually consist of 1.5 in × 3.5 in (38 mm × 89 mm) or 1.5 in × 5.5 in (38 mm × 140 mm) lumber and are commonly spaced at 16 in (410 mm) on centre. This spacing may be changed to 12 in (300 mm) or 24 in (610 mm) on centre depending on the load and the limitations imposed by the type and thickness of the wall covering used. Wider 1.5 in × 5.5 in (38 mm × 140 mm) studs may be used to provide space for more insulation. Insulation beyond that which can be accommodated within a 3.5 in (89 mm) stud space can also be provided by other means, such as rigid or semi-rigid insulation or batts between 1.5 in × 1.5 in (38 mm × 38 mm) horizontal furring strips, or rigid or semi-rigid insulation sheathing to the outside of the studs. The studs are attached to horizontal top and bottom wall plates of 1.5 in (38 mm) lumber that are the same width as the studs.

### *Interior Partitions*

Interior partitions supporting floor, ceiling or roof loads are called loadbearing walls; others are called non-loadbearing or simply partitions. Interior loadbearing walls are framed in the same way as exterior walls. Studs are usually 1.5 in × 3.5 in (38 mm × 89 mm) lumber spaced at 16 in (410 mm) on centre. This spacing may be changed to 12 in (300 mm) or 24 in (610 mm) depending on the loads supported and the type and thickness of the wall finish used.

Partitions can be built with 1.5 in × 2.5 in (38 mm × 64 mm) or 1.5 in × 3.5 in (38 mm × 89 mm) studs spaced at 16 or 24 in (400 or 600 mm) on centre depending on the type and thickness of the wall finish used. Where a partition does not contain a swinging door, 1.5 in × 3.5 in (38 mm × 89 mm) studs at 16 in (410 mm) on centre are sometimes used with the wide face of the stud parallel to the wall. This is usually done only for partitions enclosing clothes closets or cupboards to save space. Since there is no vertical load to be supported by partitions, single studs may be used at door openings. The top of the opening may be bridged with a single piece of 1.5 in (38 mm) lumber the same width as the studs. These members provide a nailing support for wall finish, door frames and trim.

### *Lintels (Headers)*

Lintels (or, headers) are the horizontal members placed over window, door and other openings to carry loads to the adjoining studs. Lintels are usually constructed of two pieces of 2 in (nominal) (38 mm) lumber separated with spacers to the width of the studs and nailed together to form a single unit. The preferable spacer material is rigid insulation. The depth of a lintel is determined by the width of the opening and vertical loads supported.

### Wall Sections

The complete wall sections are then raised and put in place, temporary braces added and the bottom plates nailed through the subfloor to the floor framing members. The braces should have their larger dimension on the vertical and should permit adjustment of the vertical position of the wall.

Once the assembled sections are plumbed, they are nailed together at the corners and intersections. A strip of polyethylene is often placed between the interior walls and the exterior wall, and above the first top plate of interior walls before the second top plate is applied to attain continuity of the air barrier when polyethylene is serving this function.

A second top plate, with joints offset at least one stud space away from the joints in the plate beneath, is then added. This second top plate usually laps the first plate at the corners and partition intersections and, when nailed in place, provides an additional tie to the framed walls. Where the second top plate does not lap the plate immediately underneath at corner and partition intersections, these may be tied with 0.036 in (0.91 mm) galvanized steel plates at least 3 in (76 mm) wide and 6 in (150 mm) long, nailed with at least three 2.5 in (64 mm) nails to each wall.

### Balloon Framing

Balloon framing is a method of wood construction – also known as "Chicago construction" in the 19th century – used primarily in Scandinavia, Canada and the United States (up until the mid-1950s). It utilises long continuous framing members (studs) that run from the sill plate to the top plate, with intermediate floor structures let into and nailed to them. Here the heights of window sills, headers and next floor height would be marked out on the studs with a storey pole. Once popular when long lumber was plentiful, balloon framing has been largely replaced by *platform framing*. It is not certain who introduced balloon framing in the United States. However, the first building using balloon framing was probably a warehouse constructed in 1832 in Chicago, Illinois, by George Washington Snow. The following year, Augustine Taylor (1796–1891) constructed St. Mary's Catholic Church in Chicago using the balloon framing method.

In the 1830s, Hoosier Solon Robinson published articles about a revolutionary new framing system, called "balloon framing" by later builders. Robinson's system called for standard 2x4 lumber, nailed together to form a sturdy, light skeleton. Builders were reluctant to adopt the new technology, however, by the 1880s, some form of 2x4 framing was standard.

Alternatively, the balloon frame has been shown to have been introduced in Missouri as much as fifty years earlier.

The name comes from a French Missouri type of construction, *maison en boulin*, *boulin* being a French term for a horizontal scaffolding support. Historians have also fabricated the following story: As Taylor was constructing his first such building, St. Mary's Church, in 1833, skilled carpenters looked on at the comparatively thin framing members, all held together with nails, and declared this method of construction to be no more substantial than a balloon. It would surely blow over in the next wind! Though the criticism proved baseless, the name stuck.

Although lumber was plentiful in 19th-century America, skilled labour was not. The advent of cheap machine-made nails, along with water-powered sawmills in the early 19th century made balloon framing highly attractive, because it did not require highly-skilled carpenters, as did the dovetail joints, mortises and tenons required by post-and-beam construction. For the first time, any farmer could build his own buildings without a time-consuming learning curve.

***Figure:** A very unusual example of balloon framing: The Jim Kaney Round Barn, Adeline, Illinois, U.S.A.*

It has been said that balloon framing populated the western United States and the western provinces of Canada. Without it, western boomtowns certainly could not have blossomed overnight. It is also a fair certainty that, by radically reducing construction costs, balloon framing improved the shelter options of poorer North Americans. For example, many 19th-century New England working neighbourhoods consist of balloon-constructed three-story apartment buildings referred to as triple deckers. The main difference between platform and balloon framing is at the floor lines. The balloon wall studs extend from the sill of the first story all the way to the top plate or end rafter of the second story. The platform-framed wall, on the other hand, is independent for each floor.

Balloon framing has several disadvantages as a construction method:

1. The creation of a path for fire to readily travel from floor to floor. This is mitigated with the use of firestops at each floor level.
2. The lack of a working platform for work on upper floors. Whereas workers can readily reach the top of the walls being erected with platform framing, balloon construction requires scaffolding to reach the tops of the walls (which are often two or three stories above the working platform).
3. The requirement for long framing members.
4. In certain larger buildings, a noticeable down-slope of floors towards central walls, caused by the differential shrinkage of the wood-framing members at the perimeter versus central walls. Larger balloon-framed buildings will have central bearing walls which are actually platform framed and thus will have horizontal sill and top plates at each floor level, plus the intervening floor joists, at these central walls. Wood will shrink much more across its grain than along the grain. Therefore, the cumulative shrinkage in the centre of such a building is considerably more than the shrinkage at the perimeter where there are many fewer horizontal members. This problem, unlike the first three, takes time to develop and become noticeable.
5. Present-day balloon framing buildings often have higher heating costs, due to the lack of insulation separating a room from its exterior walls. However, this can be remedied through the addition of insulation, as with any other framed building.

Since steel is generally more fire-resistant than wood, and steel framing members can be made to arbitrary lengths, balloon framing is growing in popularity again in light gauge steel stud construction. Balloon framing provides a more direct load path down to the foundation. Additionally, balloon framing allows more flexibility for tradesmen in that it is significantly easier to pull wire, piping and ducting without having to bore through or work around framing members.

### Platform Framing

In Canada and the United States, the most common method of light-frame construction for houses and small apartment buildings as well as other small commercial buildings is *platform framing*. In builder parlance, platform framing might also nowadays be called (only partly correctly) *'stick framing'* or *'stick construction'* as each element is built up stick by stick, which was also true in the other stick framing method, in the obsolete and labour intensive, but previously fashionable, balloon framing method, wherein the outside walls were erected, headers hung, then floor joists were inserted into a box made of walls. In contrast, in platform framing a floor box and joists making up the platform is built and placed on a supporting under structure (Sill plates, headers, or beams) where it sits flat and gets fastened down against wind lifting with galvanized metal tie straps. Once the boxed floor platform is squared, leveled and fastened then subfloor, walls, ceilings, and roof are built onto and above that initial platform, which can be repeated floor by floor, 'without the slow downs and dangers of fastening and leveling rough-sawn joists of a new floor together to the walls from ladders extending one or even two stories up.

Generally, the flooring ('platform') is constructed then the walls built on top of that layer, then another atop that, and so forth making for quick efficient labour saving construction methodologies and those have quickened further as technologies such as joist hangers have been developed to speed and enhance the technology. The methods and techniques have become so common and pervasive that even Skyscrapers use a modified form of platform framing techniques and indeed the same tools and technologies once construction builds the initial structural skeleton. Once the platform floor is laid down, the builder's crew can with chalk line, rule and pencil directly transfer an outline of the exterior and interior walls, their openings and relative locations with ease and precision from the plans or builders blue prints.

As the survey group lays down the notations and chalk lines, a carpenter crew can follow behind and lay down 2x4 'bottom plates' and tack them to the floor box. The topmost wall plates are cut only to the outside dimensions of the walls. Butting two other two by fours against these cut to size and fastened bottom plate allows the crew to rule across all three with square and lay out studs, cripple studs, and openings for that particular wall. The two loose studs are then quickly flipped on edge after openings are cut in, and studs added on the marks with quick reliable end nailing through the respective top and bottom plates. A few minutes later the whole wall section can be levered up and aligned in place and braced for later application of the top plates and adjoining walls.

The method provides builders options and flexibility such as when and where there is a floor-level opening (doorway) the next wall section can be aligned and fastened in place separately with the top plate added then used then a lintel and cripple studding added, or the entire wall could have been cut and joined at the top all along and lifted up as one entity. In the end, the outside walls are plumbed and fastened together with 'ell-configured reinforced corners' that provide nailing wood in the interior angles and strength to the building forming in effect wide posts at each corner and fastened lastly by overlapped top plates which stagger their joints from the ones capping each plate by which the studs are end nailed together. Each wall from top to bottom ends up with a doubled plate, studs, and a doubled plate, where structurally the doubled plates spread the weight of the roof and loading across the studs of the wall, ultimately to the foundation.

Overall, the framed structure sits (most commonly) atop a concrete foundation on *pressure treated wood 'sill', or 'beam'*. When on concrete, the sill plate is anchored, usually with (embedded) 'J' bolts into the concrete substrate of the foundation wall. Generally these plates must be pressure treated to keep from rotting from condensing moisture. By various standards the bottom of the sill plate is located a minimum 6 inches (150 mm) above the *finished grade* (the surrounding ground) per standard builders practices, and frequently more dependent upon building codes of the relevant jurisdiction's local building codes. In North America, building codes may differ not only state to state, but town to town, the tighter specification applying at all times. This distance, together with roofing overhangs, and other system factors, is most often selected both to prevent the sill-plate from rotting (due to the invasion of splashed water) as well as providing a termite barrier. The latter is particularly (more or less) important than anti-rotting considerations depending upon the geographical location.

Alternatively, the room, room extension, deck or even a house can be built above concrete columns U.S. builders call piers some others call pilasters, another of many term misuses common to building trade parlance. In such cases, the pier (column) is usually required to rest on bed rock or extend well below the zone of average freezing soil depth (the same as a foundation) locally, and frequently is required to also have flared out or mushroomed bottom of greater surface than that the pier top (these are called 'big foots' in the building trade, and building suppliers carry PVC molds to conserve concrete which allow a builder to satisfy area requirements and the building codes). Rigid pressure treated 'beams' (usually doubled or tripled up wider types of 2x boards) are attached to the piers using galvanized metal brackets and serve the same function as sills in foundation supported framing.

The floors, walls and roof of a framed structure are created by assembling (using nails) consistently sized framing elements of dimensional lumber (e.g. 2×4s) at regular spacings (typically divisions of 4 and 8 feet, or such as 12, 16, 19.2, or 24 inches on centre). The empty space formed between elements is called a stud bay in the wall and a joist bay in the floor or ceiling. The floors, walls and roof are typically made torsionally stable with the installation of a plywood or composite wood skin referred to as sheathing. Sheathing has very specific requirements (such as thickness and spacing of nailing). These measures allow a known amount of shear force to be resisted by the elements. Spacing the framing members properly usually allows them to align with the edges of standard sheathing. In the past, tongue and groove planks installed diagonally were used as sheathing. Occasionally, wooden or galvanized steelbraces are used instead of sheathing. There are also engineered wood panels made for shear and bracing.

The floor, or the platform in this framing type's name, is made up of joists (usually 2x6, 2×8, 2×10 or 2×12 depending on the span, on edge thus the wider joist supporting weight for a greater distance) that sit on supporting foundation walls, beams, columns or girders within and at right angle to doubled outside members also on edge (the band), forming a box. The outer perimeter is nearly the same (3 inch vs. 3.5 inches) width as the support sill. The joists will generally be installed across the shortest distance of any floor span rectangle. The outer layer of the band will overlap the inner layer with staggered end joints creating a stronger box. If joist hangers are not used, the installation of the outer board in the band is delayed to allow through-nailing directly into the ends of the joists.

The floor joists are spaced at 12 in, 16 in, and 24 in on centre, depending upon the live load needs of the design – the closer the spacing and the wider the floor joist dimension, the less the floor will flex. It is then usually covered with a 3/4-inch tongue-and-groove plywood subfloor. In the century past, 1x planks set at 45-degrees to the joists were used for the first subfloor layer, and a second layer of 1x planks set at 90-degrees to the floor cladding topped that as the second subfloor layer. In that same era, all flooring choices were a very short menu of choices between finished wood types or ceramic tiles versus today's extensive multipage menu of manufactured flooring types. Where the design calls for a framed floor, the resulting platform is where the framer will construct and stand that floor's walls (interior and exterior load bearing walls and space-dividing, non-load bearing partitions). Additional framed floors and their walls may then be erected to a general maximum of four in wood framed construction. There will be no framed floor in the case of a single-level structure with a concrete floor known as a *slab on grade*.

Stairs between floors are framed by installing three 90°-stepped *stringers* attached to wall structures and then placing the horizontal *treads* and vertical *risers* (usually about 14 of each for an 8-ft. ceiling) upon the planes formed by the stringers. A framed roof is an assembly of rafters and wall-ties supported by the top story's walls. Prefabricated and site-built trussed rafters are also used along with the more common stick framing method. *Trusses* are engineered to redistribute tension away from wall-tie members and the ceiling members. The roof members are covered with sheathing or strapping to form the roof deck for the finish roofing material. Floor joists can be engineered lumber (trussed, I-joist, etc.), conserving resources with increased rigidity and value. They are semi-custom manufactured to allow access for runs of plumbing, HVAC, etc. and *some 'common-needs' forms* are pre-manufactured as semi-mass-produced standard products made on a per order basis, like roofing trusses. Such products have a post-order lead time from several weeks to several months.

Double framing is a style of framing used in some areas to reduce heat loss and air infiltration. Two walls are built around the perimeter of the building with a small gap in between. The inner wall carries the structural load of the building and is constructed as described above. The exterior wall is not load bearing and can be constructed using lighter materials. Insulation is installed in the entire space between the outside edge of the exterior wall and the inside edge of the interior wall. The size of the gap depends upon how much insulation is desired. The vapor barrier is installed on the outside of the inner wall, rather than between the studs and drywall of a standard framed structure. This increases its effectiveness as it is not perforated by electrical and plumbing connections.

# Bibliography


Allen & Iano: *Fundamentals of Building Construction: Materials and Methods*, John Wiley & Sons Inc., Hoboken, New Jersey, 2008.

Anonymous: *"Recycling Construction Debris To Build A Park," The American City & County,* LA Public Library, Los Angeles, 2004.

Bruno, Leonard C.: *The Tradition of Technology: Landmarks of Western Technology in the Collections of the Library of Congress,* Library of Congress, Washington, 1995.

Chiras, Daniel D.: *The Natural House: A Complete Guide to Healthy, Energy-efficient, Environmental Homes*, Chelsea Green Pub., White River Junction, Vt, 2000.

Chopra, Anil K.: *Dynamics of Structures*, Prentice Hall, New Jersey, 1995.

Clough, Ray W.; Penzien, Joseph: *Dynamics of Structures*, McGraw-Hill, NY, 1993.

Condit, Carl W.: *American Building: Materials and Techniques from the First Colonial Settlements to the Present*, University of Chicago Press, Chicago, 1982.

Drysdale, D.: *An Introduction to Fire Dynamics*, New York, Wiley, 1985.

Evans, Ianto: *The Hand-sculpted House*, Chelsea Green Pub. Co., White River Junction, Vt, 2002.

Fitchen, John: *Building Construction Before Mechanization*, MIT Press, Cambridge, US, 1986.

Garrison, Ervan G.: *A History of Engineering and Technology: Artful Methods*, CRC Press, 1999, Boca Raton, Fla., 1998.

Hill, Donald Routledge: *A History of Engineering in Classical and Medieval Times,* Routledge, New York, 1996.

Ianto Evans, Michael G. Smith, and Linda Smiley: *The Hand-Sculpted House: A Practical and Philosophical Guide to Building a Cob Cottage,* Chelsea Green Publishing, Vermont, US, 2003.

Jenkins, J.C.: *The Humanure Handbook: A Guide to Composting Human Manure*, Joseph Jenkins, Inc., Grove City, PA, 2005.

Kennedy, J., Smith, M. and Wanek, C.: *The Art of Natural Building: Design, Construction, Resources,* New Society Publishers, Gabriola Island, BC., 2002.

Langmead, Donald, and Christine Garnaut: *Encyclopedia of Architectural and Engineering Feats*, ABC-CLIO, Santa Barbara, Calif, 2001.

Lehman, H. Jane: *"Builders Wake up to Recycling's Value,* Chicago LA Public Library, Los Angeles, 2004.

Lennon, T.: *Fire Safety of Concrete Structures: Backrgound to BS8110 Fire Design,* Building Research Establishment, Garston, Watford, UK., 2004.

Nawy, Edward G.: *Prestressed Concrete,* Prentice Hall, New Jersey, 1989.

Nilson, Arthur H.: *Design of Prestressed Concrete.* John Wiley & Sons, NY, 1987.

Pannell, J. P. M.: *Man the Builder: An Illustrated History of Engineering,* Thames and Hudson, London, 1977.

Paul Francis Wendt and Alan Robert Cerf: *Real Estate Investment Analysis and Taxation,* McGraw-Hill, New York, 1979.

Peters, Tom F.: *Building the Nineteenth Century,* MIT Press, Cambridge, US, 1996.

Read, R.E.H., Adams, F.C. and Cooke, G.M.E.: *Guidelines for the Construction of Fire Resisting Structural Elements,* Building Research Establishment, Garston, Watford, UK., 2004.

Sherman, Rhonda: *The Inside Story of the Greenest Building Complex In the U.S.,* LA Public Library, Los Angeles, 2004.

Smith, Michael G.: *"Cob Building, Ancient and Modern," in Kennedy,* Smith and Wanek, 2002.

Thompson, Brian L.: *Builders Look To Cut Job-Site Waste,* LA Public Library, Los Angeles, 2004.

Vale, Brenda and Robert: *The New Autonomous House,* Thames & Hudson Ltd., London, 2000.

Willis, Carol: *Form Follows Finance: Skyscrapers and Skylines in New York and Chicago,* Princeton Architectural Press, NJ, 1995.

Wisher, R.W. and Smart, P.M.T.: *Results of Fire Resistance Tests on Elements of Building Construction,* Building Research Establishment, Garston, Watford, UK., 1975.

Woolley T.: *Natural Building: A Guide to Materials and Techniques,* Crowood Press, Marlborough, Wiltshire, UK, 2006.

# Index

## H



## I



## J



## L



## M



## P



## R



## S



## T



## V



## W



❑❑❑